Chart and Graph Preparation Skills

Chart and Graph Preparation Skills

Tom Cardamone

Illustrated by
Tom Cardamone and Ann Kahaner

VNR VAN NOSTRAND REINHOLD COMPANY
New York Cincinnati Toronto London Melbourne

Library of Congress Catalog Card Number 80-26207

ISBN 0-442-26284-1 (cloth)
ISBN 0-442-26286-8 (paper)

Printed in the United States of America

Published by Van Nostrand Reinhold Company
135 West 50th Street, New York, NY 10020

Van Nostrand Reinhold Limited
1410 Birchmount Road
Scarborough, Ontario M1P 2E7, Canada

Van Nostrand Reinhold Australia Pty. Ltd.
17 Queen Street
Mitcham, Victoria 3132, Australia

Van Nostrand Reinhold Company Limited
Molly Millars Lane
Wokingham, Berkshire, England

16 15 14 13 12 11 10 9 8 7 6 5 4 3 2 1

Library of Congress Cataloging in Publication Data

Cardamone, Tom.
Chart and graph preparation skills.

Includes index.
1. Graphic methods. I. Title.
QA90.C29 001.4'226 80-26207
ISBN 0-442-26284-1
ISBN 0-442-26286-8 (pbk.)

Table of Contents

To my daughter
Doreen Lynn

INTRODUCTION

Long before man could write words, he made "signs." At first he probably only scratched markings on a tree or scribed personalized lines on a rock to identify a location. As time passed and his awareness of life around him increased, so did his urge to document his existence. He painstakingly etched symbols into rock, surely in anticipation of communicating his thoughts to someone, sometime. Eventually these simple pictures, which became more sophisticated with his intellectual growth, developed into quite an elaborate method of more precise communication. If early man could only see us now!

Graphic impressions, the oldest method of direct recording and communication, are still used today. In fact, the use of graphs or charts is in demand more now than ever before. Our accumulation of knowledge during the past few decades has accelerated with such startling velocity, it is a wonder we can keep up with ourselves. Because this mind-boggling, complex mass of information must be made understandable, recorded, and conveyed, we have invented tools, machines, and other devices to help us. One of the most useful tools that has survived time is the chart or graph.

Today, of course, we do much more than just scratch lines or draw pictures. The modern-day graph or chart has evolved into an invaluable vehicle of detailed information. Because graphs and charts are used so widely, it is literally impossible to categorize their application. They are used to illustrate anything and everything from a simple comparison between two vegetables, to a vivid description of a massive business operation, or an intricate delineation of the beats of an electrical pulse. In fact, I cannot visualize today's world functioning efficiently without the use of graphs in one form or another. One doesn't have to be a scientist or a mathematician to use or construct a graph. For example, when you keep score at a bowling match, outline a food budget, plan a trip, develop a diet, or schedule your semester classes, you're using a form of chart or graph.

While most people are familiar with graphs in simple form, they are not fully aware of their potential usefulness. The more one understands how to approach the making of a plan, program, or project, the better its chance of success.

In discussions about graphs, the emphasis is usually placed upon familiarization: how to read them, their various types, and their application. Most people understand that phase, but how many know how to *prepare* a graph?

Granted, some graphs can be quite intricate, requiring a highly developed skill, but there are also less complicated forms that are not difficult to execute. Simple charts such as those used in annual reports are extremely common and relatively easy to prepare.

A sales meeting is hardly complete without the display of some form of chart. A hospital without charts would be hardpressed to function efficiently. Not only does our present age of electronics demand that we use charts and graphs, they are becoming more complex. That is why it is so important to know how to prepare them. The most encouraging aspect of their preparation is that no special skill is required. All that is necessary is a bit of patience, some simple logic, and a bit of basic arithmetic. One does not have to possess a particularly deep understanding of the subject being charted. Of course, the more technically involved a chart is to be, the more demanding it becomes to execute, but the execution of everyday, fundamental graphs can be learned in a relatively short time.

At first glance the step-by-step procedure illustrated here may leave the impression that it is quite complicated. On the contrary, if you follow the carefully organized directions, you will be amazed at how simple charts actually are to make. Because the directions are broken down to refined details, they may appear to represent a much longer task, but it is no different than watching a baseball pitcher in slow motion. The actual action takes a few seconds. Here, four or five pages of illustrations may represent only four or five minutes work.

This book presents a basic understanding of the structure of graphs and step-by-step directions for preparing them on a professional level.

Bear in mind that a graph can be prepared with as much precision as is necessary. Absolute hairline accuracy may not always be needed. Should precision be required, it cannot be faked. That is why it is best to learn this skill as efficiently as possible. It can be used either as a basis for a lucrative professional career or simply as an added tool in everyday life. At one time or another, most of us will have to prepare some kind of chart to explain something. This book will unquestionably help you do it better.

Chapter One
Understanding Charts

The general purpose of a chart or graph is to give a visual comparison between two or more things. For example, an event of one year may be compared with a corresponding event of another year, or the population of one location compared to the population of another location, or perhaps the changes in budget from one year to the next may be represented. A graph might show the ratio of a part to the whole, such as the percentage of all American automobiles manufactured in 1970 that had stick shifts.

One significant reason for visualizing a comparison is to reinforce its comprehension. Although it is easy enough to realize the impact of something one reads or hears such as, "Tom Francis earned $4,200 during the year 1950 and $39,000 in 1975," a graphic interpretation of the same facts could be more effective, as shown in Fig. 1.

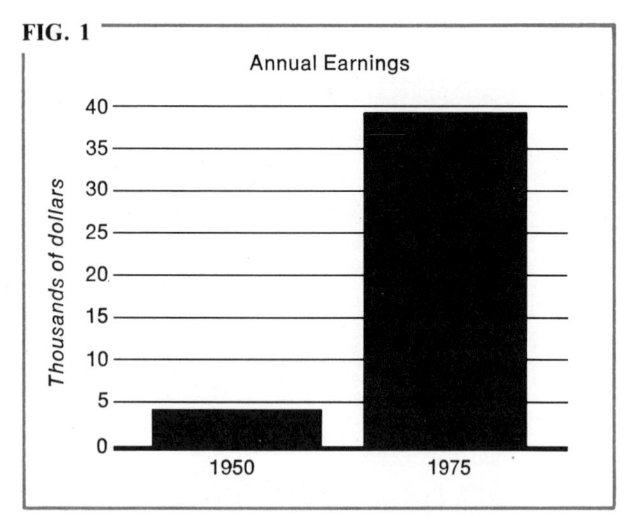

FIG. 1

Charts, graphs, and numerous devices and configurations are used to dramatize a statement, a fact, a point of view, an occurrence, an idea, etc. In today's scientifically oriented world there is an increasing demand for visual aids to assist in the quick comprehension of both simple and complex statistics or problems.

It takes longer to digest the meaning of an itemization of compiled figures than if the same figures are presented graphically. In Fig. 2, A and B describe identical facts. The tabular chart A requires a fair amount of study in order to grasp the full meaning of the figures, but in B not only is comprehension faster, another dimension has been added to its meaning.

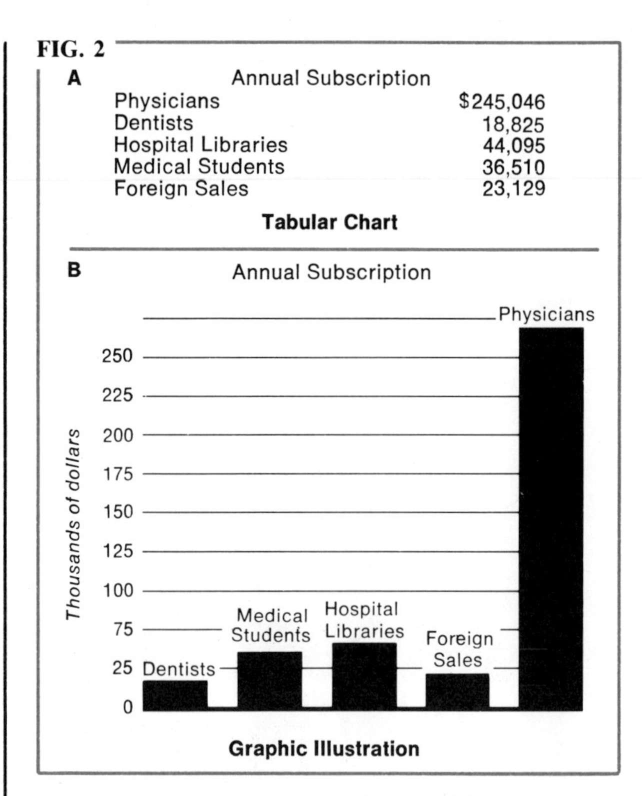
FIG. 2

A

Annual Subscription	
Physicians	$245,046
Dentists	18,825
Hospital Libraries	44,095
Medical Students	36,510
Foreign Sales	23,129

Tabular Chart

Graphic Illustration

There are no set rules for graphic interpretation, but there are certain basic principles and formulas one must understand in order to develop such graphic designs. It is no different from learning to use a hammer, saw, and a few other tools before building a house—you'll simply build a better house.

When working with charts you must sharpen your imagination along with your pencils. It requires the use of simple mathematics combined with basic geometric projection methods. You'll have to think creatively while simultaneously adhering to rigid formulations.

Because some of the terminology of this subject is redundant, I will try to simplify it as we progress. To start with I will refer to all types of charts and graphs as *charts* unless otherwise indicated.

To start with, let's look at the general categories of standard charts. Probably the most commonly used is the *column chart* (Fig. 3). Besides being effective for demonstrating a comparison between two or more things, it also offers much opportunity for design variations (see Chapter Seven).

FIG. 3

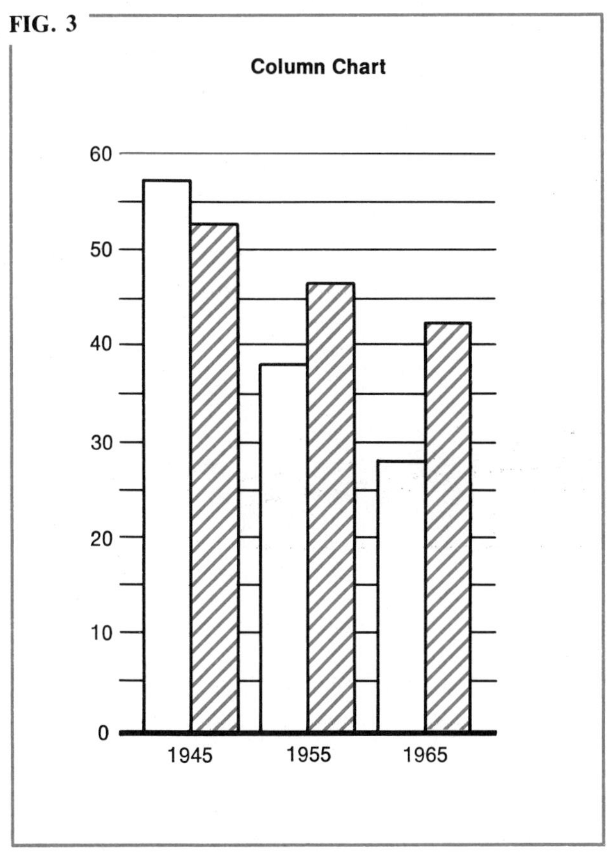

Next is the *bar chart,* which is essentially a column chart on its side and is used for the same purpose (Fig. 4). Although the column chart is sometimes also referred to as a bar chart, it is best to call them by their proper names. This helps differentiate the vertical from the horizontal type.

FIG. 4

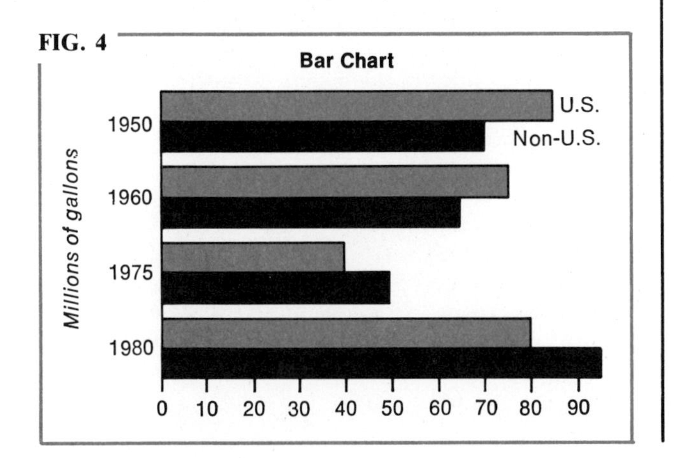

Of course, the ever-popular *pie chart* also lends itself very nicely to a variety of graphic designs (Fig. 5). It is usually used when representing percentages of a whole (100%).

FIG. 5

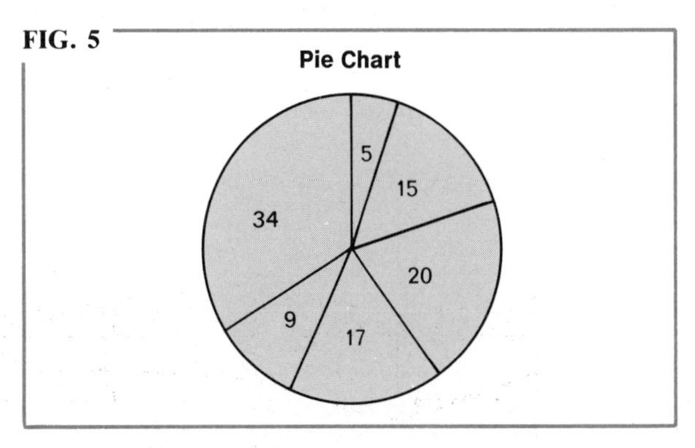

In Fig. 6 we have a *line chart,* more commonly called a *line graph,* which in itself could create a bit of confusion.

The line chart is exceptionally impressive when comparing several things but could present a visual problem if the comparisons are too many or too close in relation to one another. If not used properly, it could easily confuse the reader, thus defeating the purpose of the chart.

FIG. 6

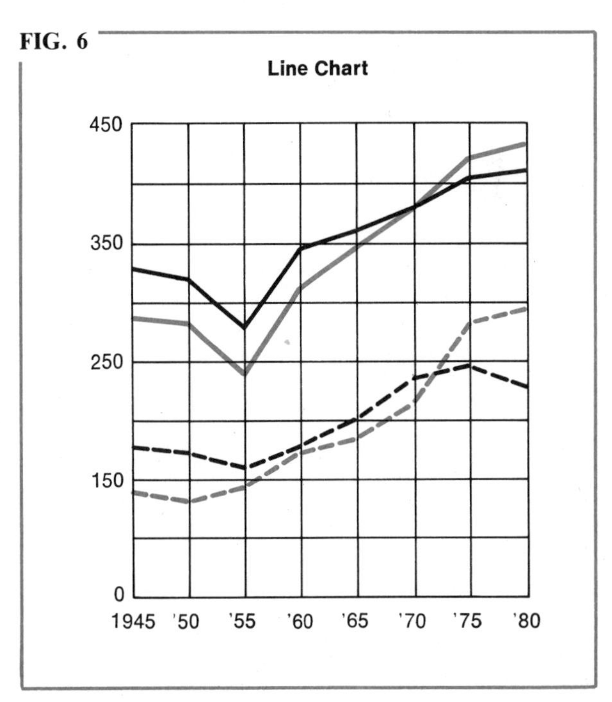

A *layer chart* (Fig. 7) is linear in appearance but has a different representation. It depicts the accumulation of individual facts stacked one over the other to create the overall total. This type of chart is a bit more complex than the others, since it illustrates much more. In addition to showing the comparison of layers that add up to the total, this type of chart also shows how each group of layers relates to subsequent groups.

FIG. 7

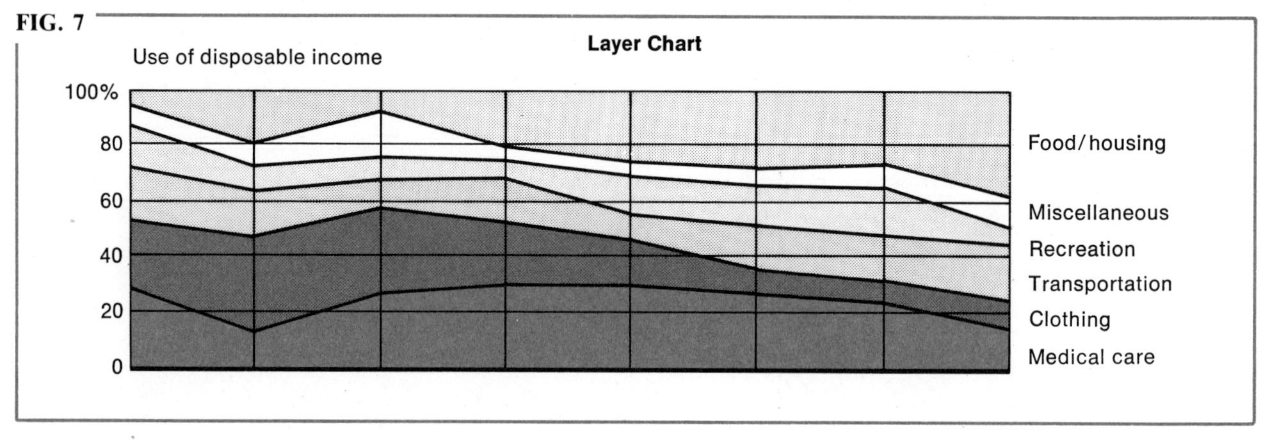

THE SCALE

The principles by which each of these five types of charts is constructed is the foundation of most other, more intricate types of charts. It would be best to concentrate on thoroughly understanding the five basic charts before going on to other, more imaginative variations.

Although each chart looks different and may be used for a different purpose, they all have one thing in common—a scale.

By a *scale* we mean a basis for progressive series in accordance with a set ratio such as a ruler. A 12″ ruler is a scale marked off in inches, half inches, quarter inches, eighth inches and sixteenth inches (Fig. 8). With a little imagination these subdivisions can be used in many ways to represent much more than measurements on a straight line, as you will see further on.

FIG. 8

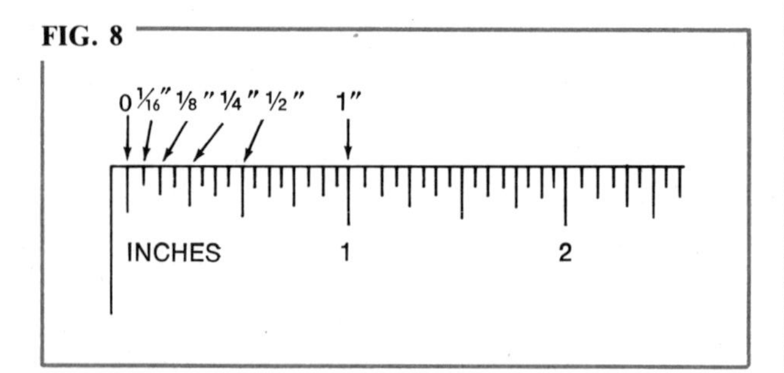

The basic structure of a chart is quite simple. You will note the common "grid" or crossed horizontal and vertical lines illustrated in Fig. 9. The vertical lines generally represent a progression of time, e.g., year to year or month to month. The horizontal lines are usually a measurement of some kind depending on the nature of the chart, such as number of dollars or number of people. For now let's concentrate on the structural function.

The distance between the horizontal grid lines in the chart itself is a fixed or set ratio representing whatever the particular chart is meant to illustrate. In Fig. 9 the chart represents dollar figures, clearly identified along the side. In this chart the space between *each* pair of adjacent horizontal lines is calculated to represent exactly the same number of dollars, namely, $10. The bottom, heavier line is called the *base* line and generally represents zero.

The first (left) vertical line usually carries the increment identification of the horizontal grid lines and is called a *scale*. Often the horizontal lines extend slightly beyond the vertical line for easier readability (this is only a matter of style). These short extensions of the lines are called *tick marks*.

FIG. 9

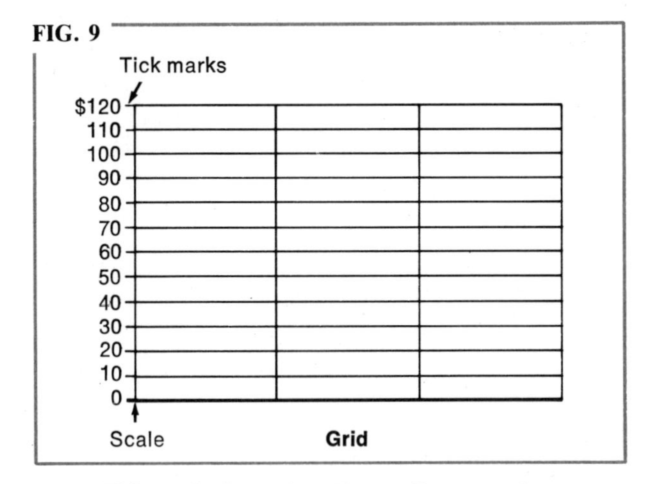

This scale line (though smaller) may be compared with the one-inch marks along the edge of a ruler. On the ruler the marks represent inch increments, and on the chart they represent $10 increments. We have actually created our own "ruler" for measuring dollars. The length of the scale line and the space between tick marks is arbitrary or at least determined by the size of the chart you are drawing. The important thing is that the space between tick marks must be accurately uniform. For example, in Fig. 10 charts A and B illustrate the same dollar figures. The only difference is their size. Although B is larger than A, the scales are identical in what they represent; the overall height and number of tick marks runs from 0 to $120 on both scales.

FIG. 10

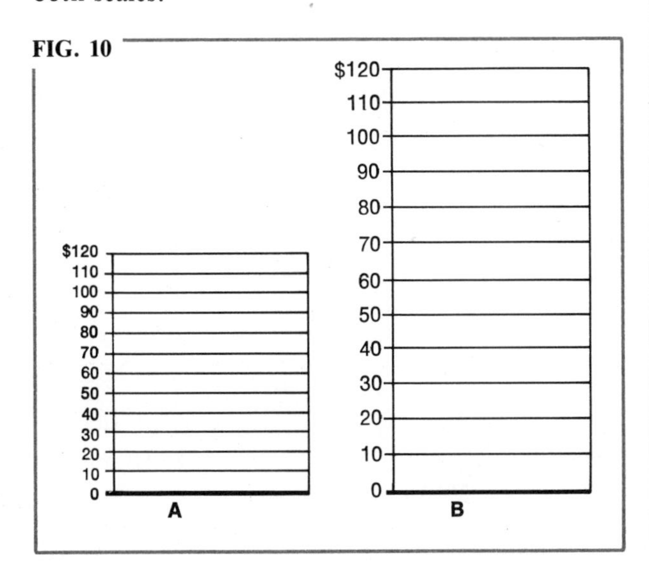

However, a ruler is a *fixed* scale; that is, the marks on rulers of any size measure exactly the same and have the same meaning. But the scales used in charts are flexible, since they can be drawn to any length necessary to accommodate the size of the chart being constructed. Furthermore, a chart scale may be subdivided into as many increments as required.

In Fig. 11A the scale represents actual dollars from 0 to $120 and is broken down into $10 increments. In B the overall scale represents the same dollar figures but is broken down into $5 increments.

There is no set rule for a scale subdivision. It is purely up to the discretion of the "chartist" and the requirements of the chart.

FIG. 11

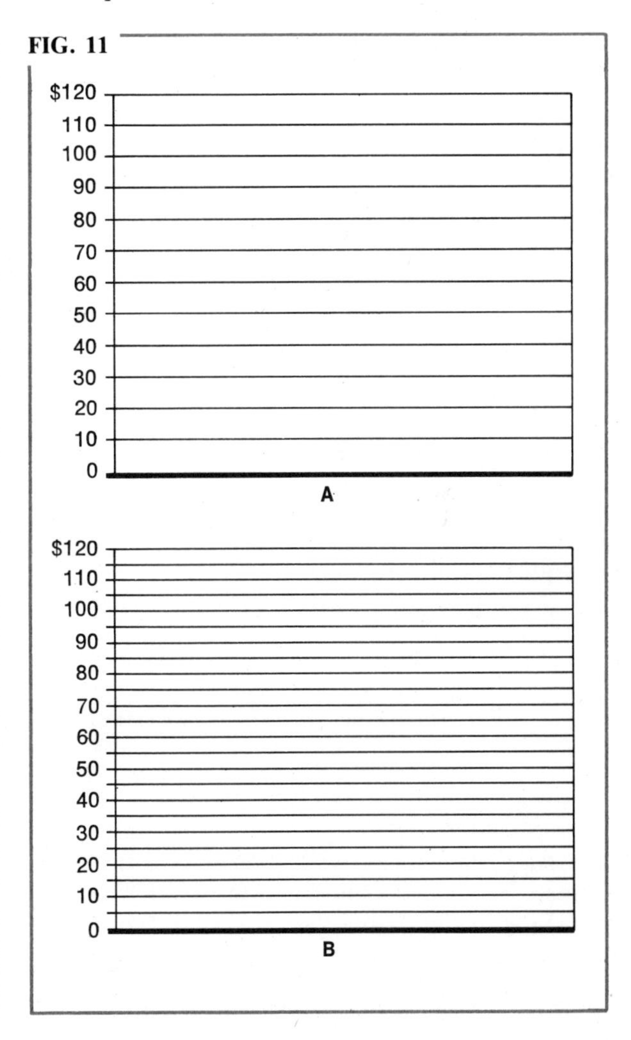

Now that you have been introduced to the scale, it is time to begin using your imagination. In Fig. 12A the chart illustrates the actual number of dollars, the total being $120. In B it shows the actual number of people, the total being 120 people. In C it represents thousands of cars, the total being 120,000 cars. In D it represents millions of people, the total being 120,000,000 people.

FIG. 12

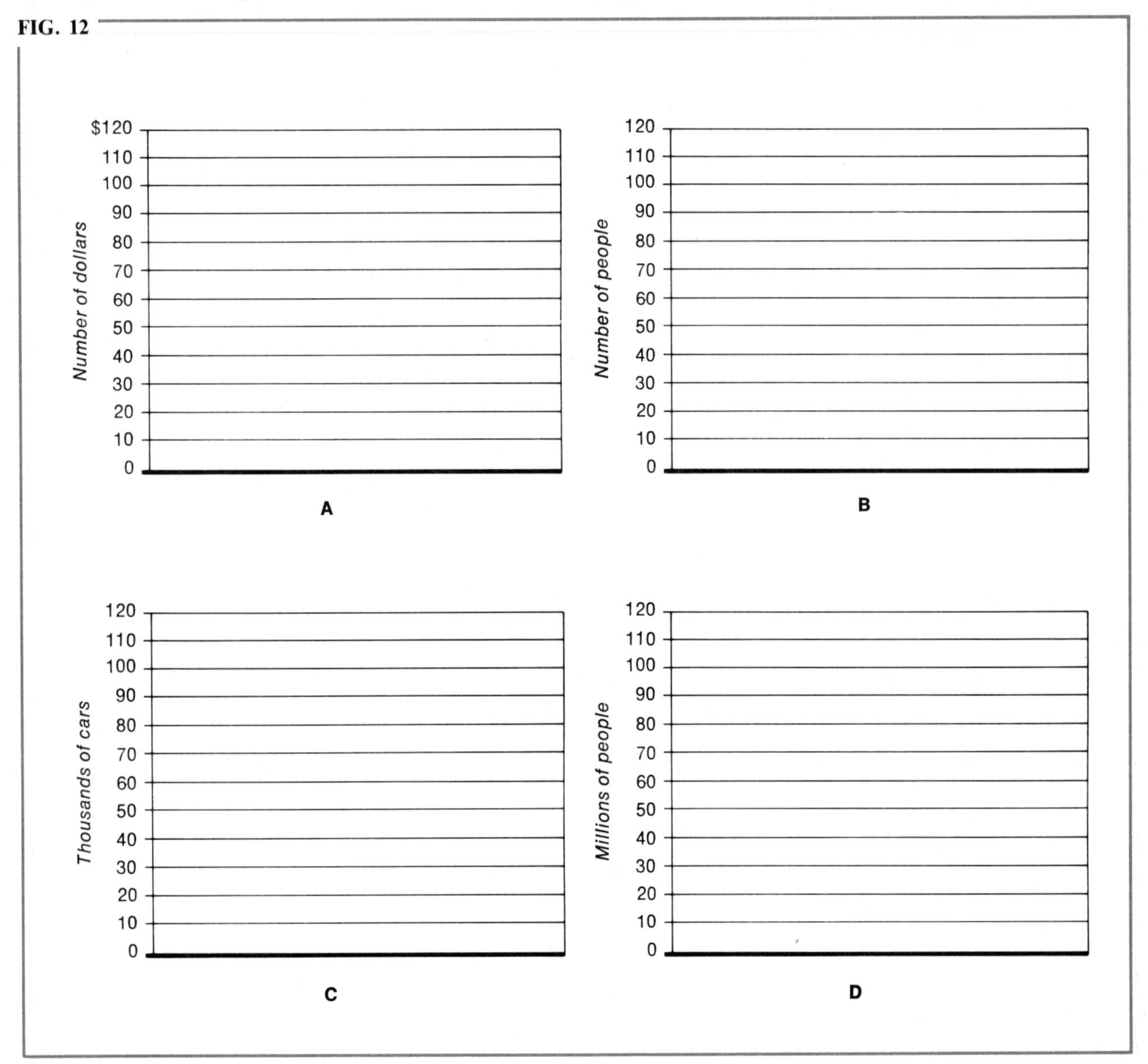

Note that, except for the identification or meaning of the scale, all four charts are identical.

What is the significance of this? It simply means that a scale in itself has no fixed meaning. It can represent whatever you choose in any quantity that is practical for the purpose of that particular chart.

In summary, a scale can be any length, subdivided into as many equal increments as you choose, representing any quantity, and can mean anything you desire.

HOW TO CONSTRUCT A SCALE

There are numerous methods used for the construction of a scale, some of which are simple, while others are a bit more complicated and perhaps unnecessary. Regardless of which method is used, the principles are the same. The methods I will describe seem to be the most effective. Take particular notice of the fact that very little arithmetic is used in their application. They are primarily geometric projections mixed with a trick or two. Do not confuse a trick of the trade with a gimmick. Gimmicks are for the incompetent; tricks are for the professional. At no time will quality be sacrificed. You are studying a skill. It takes time, ingenuity, practice, and above all, an enormous amount of patience.

You will have to tax your imagination and exercise some creative thinking at this time. Always bear in mind that there is no set meaning to a chart until you determine what it should be. Perhaps the best place to start is with the common ruler. For example, if you were to draw a scale 6″ high, you could use your ruler to measure off 6″. This scale could represent a total of 6,000 cows, with the tick marks at every inch for 1,000 cows (Fig. 13).

What you have actually done here is change the meaning of the ruler. The ruler itself represents numbers of inches. You marked off *6″*, but this length is meant to represent *6,000 cows.* Then you marked off each *1″* mark but changed its meaning to *1,000 cows.*

When you construct a scale, you are really creating your own personalized, tailor-made ruler. For example, the chart just described could alternatively be done 3″ high to represent *6,000 cows* and every $\frac{1}{2}$″ mark ticked off for each *1,000 cows.* Or it could be drawn $4\frac{1}{2}$″ high, and every $\frac{3}{4}$″ mark would represent 1,000 cows, and so on. The idea is to use the divisions on the ruler any way you can to create the scale you need.

So by counting off any number of equal increments on the ruler (such as so many half inches, or quarter inches, or eighth inches) you create a scale, simultaneously establishing the total length of the line. This is an excellent method when the total length of the scale is not critical, that is, when the space between tick marks is the critical measurement rather than the total length of the scale. When all the distances between tick marks

FIG. 13

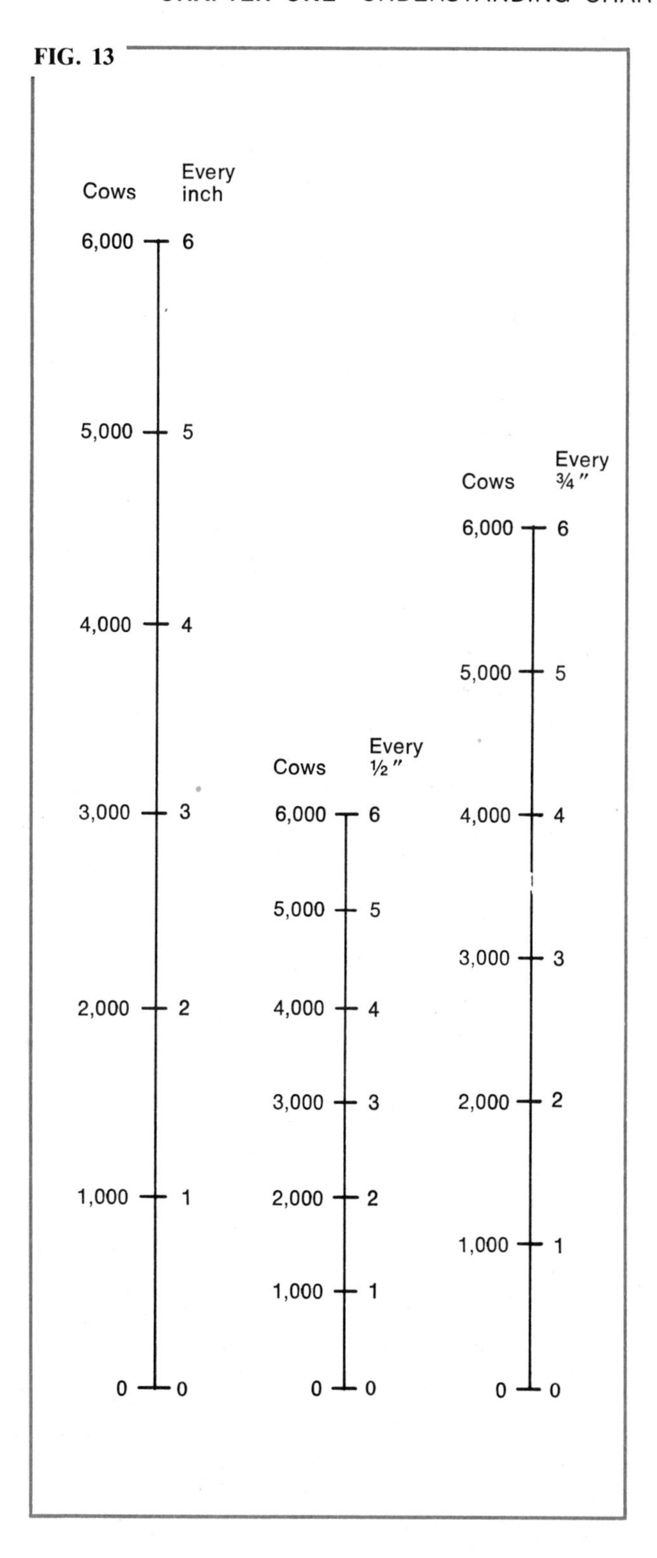

are totaled, the length of the scale is automatically established. But it is not always this simple. For the most part, you will be working within a *given* area, having first established the overall length of the scale. The distance between tick marks is to be calculated within that given length. This may not be easy to work out with a ruler. Suppose the established length of a line is one that could not so easily be subdivided with a ruler. Remember, while your answer can be worked out mathematically, you must also convert it to the divisions on a ruler. A line 5 $\frac{5}{8}''$ long divided into 12 equal parts could present a problem. Mathematically 12 divided into 5 $\frac{5}{8}''$ equals $\frac{15}{32}''$. Now ticking off every $\frac{15}{32}''$ on your ruler could be done, but it would be a chore. Suppose the line measured 5 $\frac{11}{16}''$. When divided by 12 this would equal 7.58 sixteenths of an inch or 15.16 thirty-seconds of an inch. Find *that* on your ruler. It would be difficult to do with any degree of accuracy. To carry this example further, how would you break down 5 $\frac{11}{16}''$ into 17 equal parts? Or 31 equal parts? Using arithmetic and a ruler for this kind of problem is quite awkward.

My purpose is to outline the preparation of accurate charts, executed with professional, hairline precision. Keep in mind that if you can easily define an increment on your ruler that subdivides a given line into the number of parts you require, this approach is perfectly acceptable. The problems arise when you *cannot* find equal increments on your ruler to work with.

Now is the time to explain the first trick of the trade. Look back at our answer of 15.16 thirty-seconds of an inch. Laboriously counting off this length twelve times is ludicrous and in fact probably impossible. What you might try is to tick off a single length of 15.16 thirty-seconds of an inch (if you can) along the straight edge of a piece of paper. Using this edge of paper as your ruler, you can tick off 12 to get your total (Fig. 14).

This will only work if your tick marks are *exactly* 15.16 thirty-seconds of an inch. This trick may not be the answer for this problem. But remember it; it will be very useful later on.

Another method is to use a tool called *dividers* for measuring and transferring dimensions. This is done frequently but has decided disadvantages. You must have good quality dividers and be extremely careful not to damage the surface of your drawing paper with the sharp points. Whether you use dividers or tick marks

FIG. 14

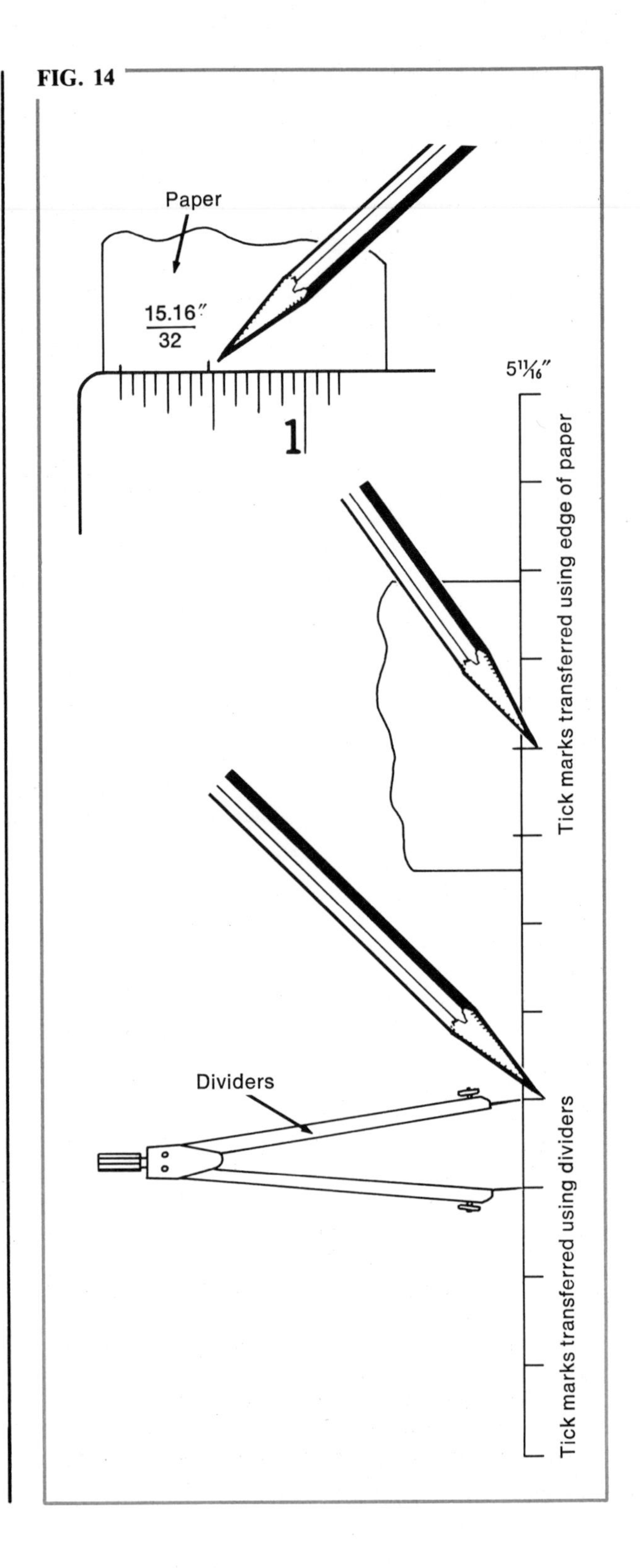

along the edge of a piece of paper, you still have to establish the distance in question on your ruler. Of the two methods, the one employing tick marks is the more reliable. It all depends on how you use it. I will get to that later.

The point here is that working with odd dimensions and fractions presents a problem. In order to construct an accurate chart, you must eliminate the guesswork. Nothing is to be done by eye (commonly referred to as "eyeballing it"). Now is the time to discard the mathematics and turn to something more accurate—geometric drawing.

HOW TO SUBDIVIDE A GIVEN LINE

Arithmetic and mathematics take a back seat with this method. Actual figures and measurements are to be used sparingly or only when indicated by the requirements of the project.

It is advisable that as you read these step-by-step methods, you actually carry them out on a piece of ledger bond paper. Before you start, let's talk about your tools. Constructing charts is a no-nonsense business.

You *must* work with quality tools: no thick, plastic-edged T-squares or heavy, wooden grade-school rulers.

And while we are on the subject of rulers—despite what you may have been told in school or have read in art books—*never* measure from the 1″ mark on your ruler. Always measure from the 0 mark (the first tick mark). Measuring from the 1″ mark is a common but most unprofessional way to read a ruler. First of all, it stems from using the old-fashioned wooden rulers with thick tick marks that start at the absolute edge. Of course you would have trouble with that type of ruler. You just cannot do professional work with it. You must use a thin, quality-grade steel ruler or a good ivory-edged wooden ruler. If you have already developed the habit of measuring from the 1″ mark, start now to break it. One day you'll spend three hours constructing something that should be 11″ long and discover you forgot to "add the inch."

In addition to good tools, your pencil is of utmost importance. When you have gained skill, you will use a 9H pencil, sharpened to a needle point, but while you are learning, it is best to work with 6H. Graduate to a 9H once you have learned to handle and respect such a hard pencil. By using it improperly you could destroy a job. There are only two or three do's and don'ts regarding the use of your pencil. The most vital "don't" is *never* spin or press a tiny hole in the board with the point of your pencil while measuring. This is the most disastrous habit among beginners. You cannot erase the hole, and you cannot draw a proper ink line over it or from it or up to it.

Another "don't" is do not draw back and forth along a straight edge or ruler while penciling a line. Also, do not "push" the pencil toward the point while drawing a line. Hold your pencil as perpendicular to the surface as possible and firmly (with just enough pressure to leave a clean, light line) guide it along the edge in one direction only. Too much pressure or too many back and forth strokes will tear the surface of your drawing paper and ruin the surface.

Now, working with the good tools and a hard pencil sharpened to a fine point, you are ready for the first lesson.

Tape down a single sheet of clean, white ledger bond paper. Do not do this while the sheet is still in the pad. Now, with a T-square and triangle construct a vertical line $5\frac{11}{16}$″ high. Do not draw too close to the edge of your paper. Give yourself some room to work with. If you are using an 11″ by 14″ sheet of paper, you could draw this line approximately 3″ from the left side and 3″ down from the top. Remember, do not press too hard—hard enough to see the pencil line but not to tear the surface of your paper. It will not take long to learn the proper pressure while working with hard pencils (Fig. 15).

FIG. 15

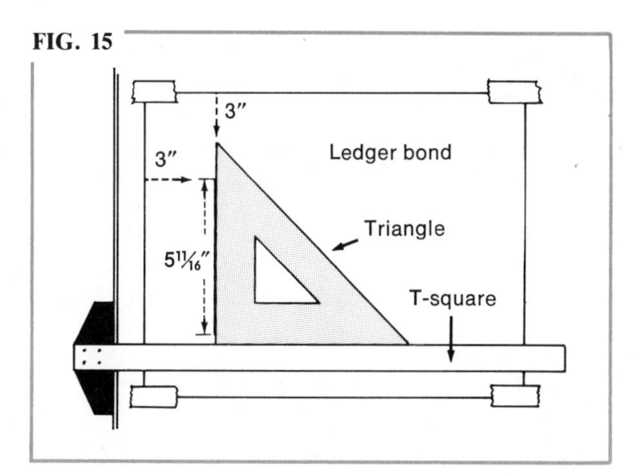

Now draw a horizontal line about 7″ long at a right angle to the bottom of the line (Fig. 16). The problem: divide the vertical line into 12 equal parts.

Use a little mental arithmetic here. Look at your ruler and find 12 equal increments that are easy to define. The most obvious would be the 12 one-inch marks or perhaps every $\frac{1}{2}$ inch mark. They are easy enough to find, and 12 half inches will equal 6 inches. You can work with that (Fig.16).

FIG. 16

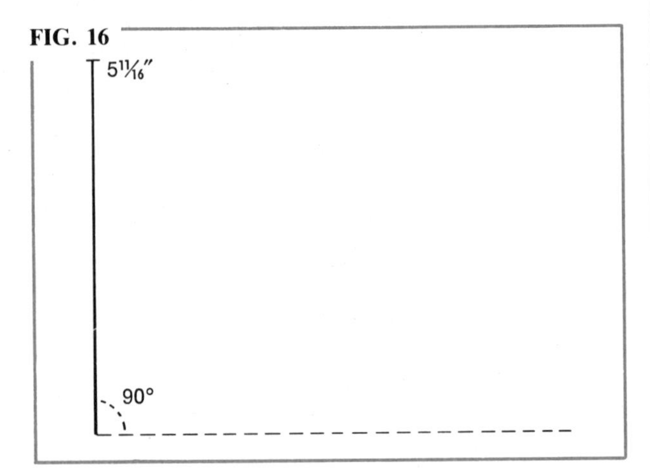

Place your ruler along the horizontal line, aligning the 0 mark with the vertical line, and very carefully and as *accurately* as possible tick off 12 half inch marks. To repeat, do not press a hole while marking. Just a simple, clean dot or short line is all you need (Fig. 17).

FIG. 17

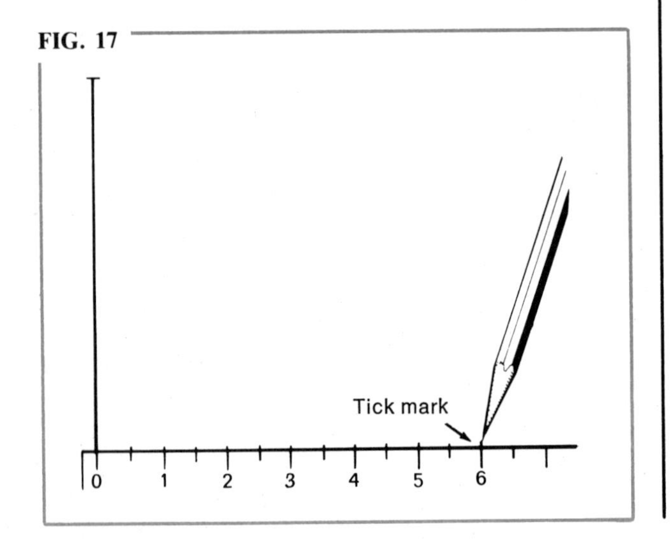

This next step can be done in either of two ways. One is to use an adjustable triangle. Place it on your T-square and adjust the edge until it touches the top mark, $5\frac{11}{16}$″ (A), on the vertical line and the twelfth mark (B) on the horizontal line (Fig. 18A). Lock it in this position and move the triangle along the T-square until it is perfectly lined up with the next to last (eleventh) tick mark (C) on the horizontal line. Now draw a light cross mark on the vertical line (D) where it is intersected by the angle of the triangle (Fig. 18B).

FIG. 18

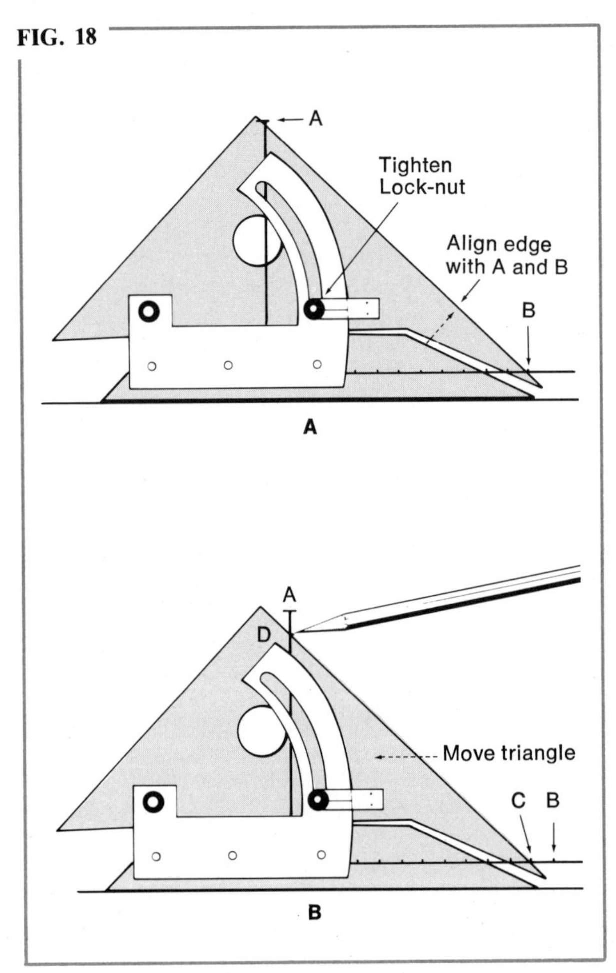

Continue to move the triangle along the T-square, lining it up with each tick mark on the horizontal and ticking off each intersection on the vertical line until all 12 marks have been transferred from the horizontal line to the vertical. If you do this with great care and accuracy, each of the 12 tick marks on the vertical line should be exactly the same distance apart. Thus you have accurately divided your line into 12 equal parts (Fig. 19) without using mathematics and without even knowing what the distance between marks actually measures. You don't need to know.

FIG. 19

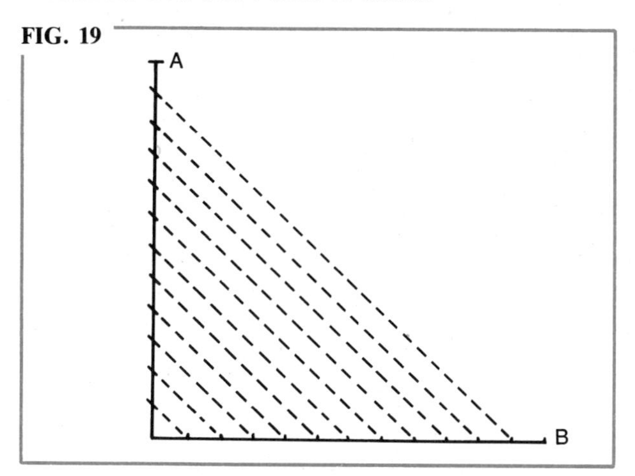

To double check the accuracy of your work, tick off the distance between two of the marks along the edge of a piece of paper (as in Fig. 14) and check it against the other marks you have drawn. They *should* all line up with your marks on the edge of the paper.

If you do not have an adjustable triangle, untape the drawing paper and turn it until points A and B are in line with your T-square. Tape the paper in this position and tick off the marks on the vertical line using your T-square (Fig. 20).

FIG. 20

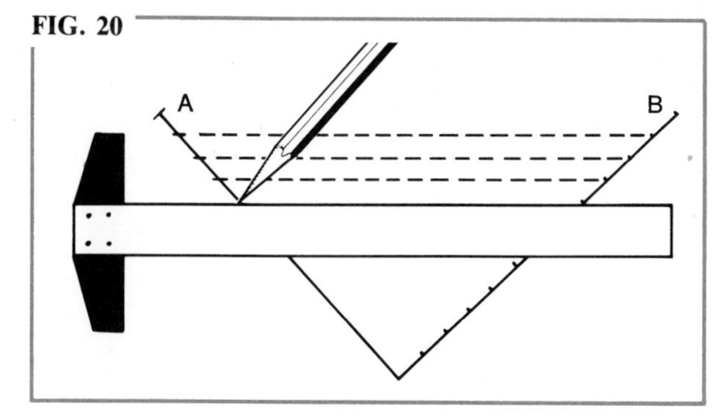

While this is an effective method, it isn't always possible. The board or paper you're drawing on, in addition to the size of the chart, may present a problem.

This method should be avoided when drawing charts for the simple reason that once you've started the drawing it should not be removed until it is complete. You want nothing to interfere with the accuracy of your drawing. Removing it before it is finished makes it difficult to realign the drawing exactly the way it was when you started. This could result in an inaccurate drawing.

Rather than using an adjustable triangle or turning the artwork to align with your T-square, here is another approach. You may find this more practical. Try it.

First construct the $5\frac{11}{16}''$ vertical. Then lightly draw a horizontal line across from the $5\frac{11}{16}''$ mark (Fig. 21).

FIG. 21

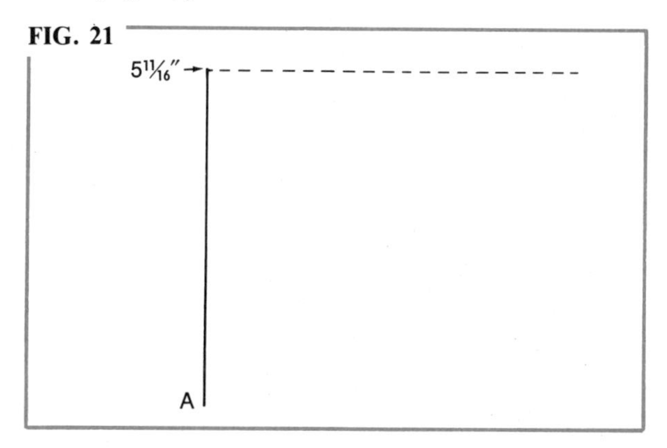

From here on, we will refer to the beginning of the ruler as the 0 mark.

Now, align your ruler 0 mark with the bottom (A) of the vertical line and pivot your ruler until the 6″ mark (B) touches the top horizontal line (Fig. 22).

FIG. 22

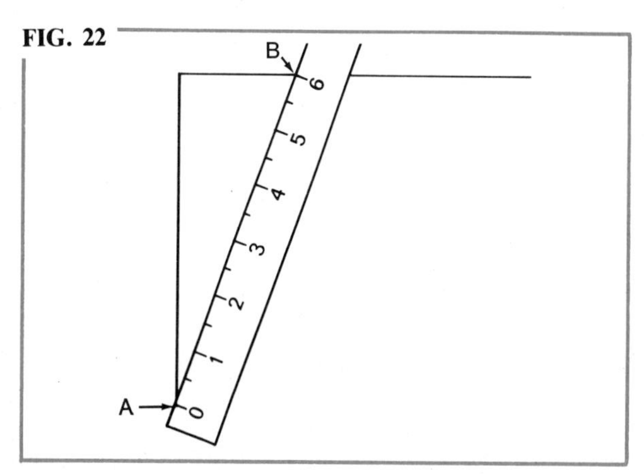

Hold your ruler firmly in this position and draw a light guide line along its edge from 0 to 6″. This guide line is only a safety device for the purpose of checking your accuracy as you work. Should you accidentally shift the ruler, even slightly, it will be noticeable. Then tick off each $\frac{1}{2}$″ mark along this guide line (Fig. 23).

FIG. 23

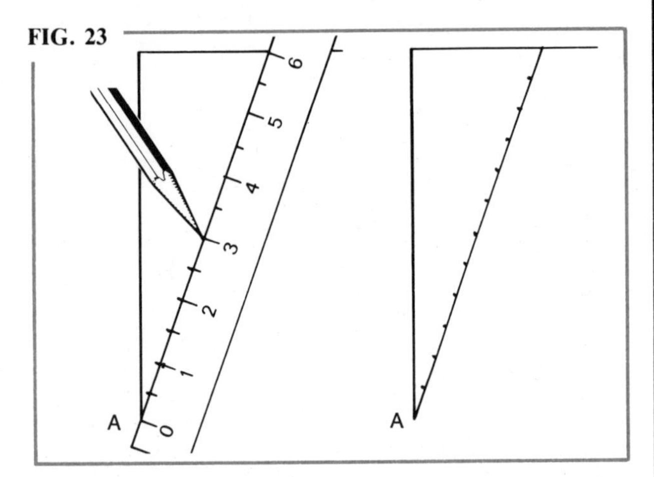

Remove the ruler, line up your T-square with each tick mark, and transfer the marks to the vertical line (Fig. 24). Simple?

FIG. 24

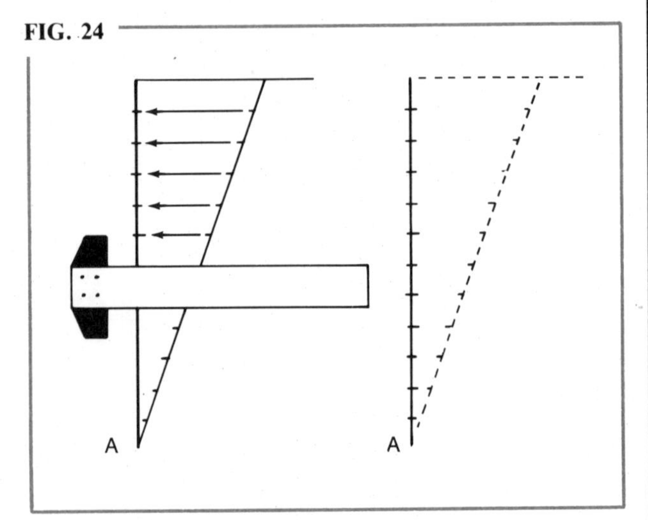

You have just created a scale $5\frac{11}{16}$″ high with 12 subdivisions. Just to be certain you understand why these approaches are more accurate and probably easier than using mathematics, tick off the space between two of the marks on the vertical line (onto the straight edge of a piece of paper) and compare the marks with the $\frac{1}{2}$″ mark on your ruler. It should be just a hair wider than $\frac{7}{16}$″. 7.58 sixteenths, perhaps? Who knows? It does not matter, as long as it is correctly drawn and you have 12 equal parts.

You have just created your first scale.

Let's take it one step further. Change the requirements from 12 equal parts to 15 equal parts. What should you do? Just use your imagination a bit. Find fifteen $\frac{1}{2}$″ marks on your ruler (total $7\frac{1}{2}$″). Repeat all the steps in either method just described, but instead of using the 6″ mark you will use the $7\frac{1}{2}$″ mark and tick off every $\frac{1}{2}$″ (Fig. 25).

FIG. 25

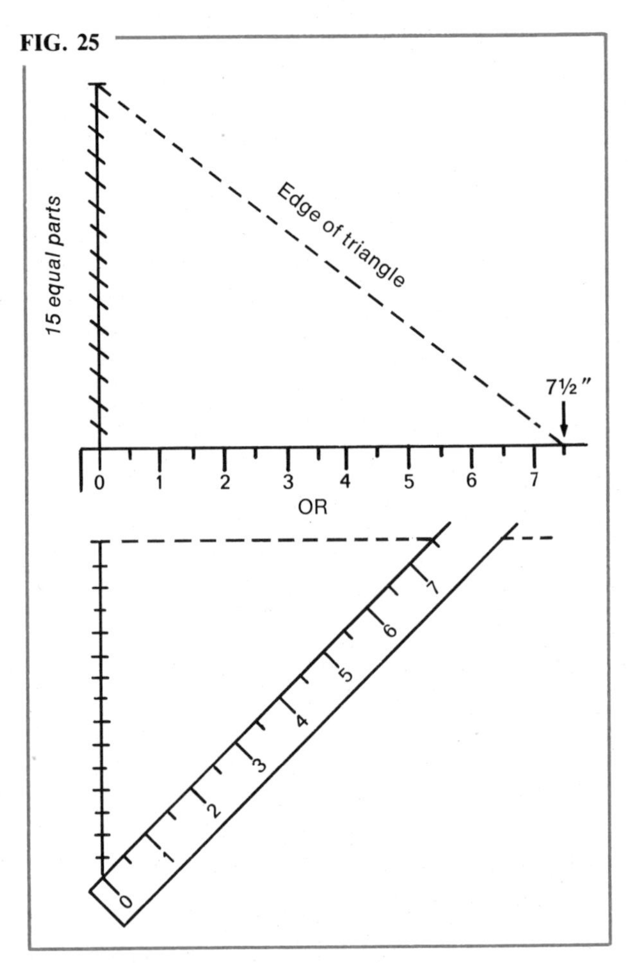

How about 37 equal parts?

Find any 37 equal segments on your ruler and use that particular measure. For instance thirty-seven $\frac{1}{4}$″ marks measure $9\frac{1}{4}$″. Use $9\frac{1}{4}$″ and mark off every $\frac{1}{4}$″ (Fig. 26).

FIG. 26

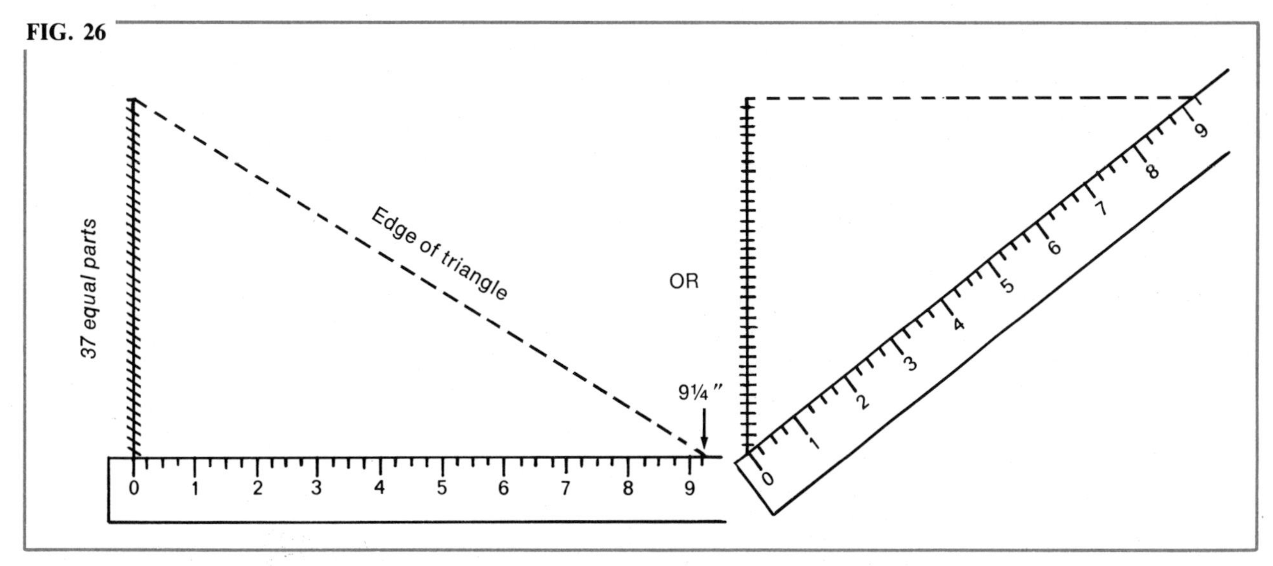

Suppose you only needed 5 equal parts? If you were to use the first method (Fig. 18), you would simply measure 5″ and tick off every inch. But if you use the second method (Fig. 22), you will find that the 5″ mark does not reach the top horizontal line. Of course not. The line is $5\frac{11}{16}$″ long to begin with. What should you do now? Use a little mental arithmetic. Why not use the 10″ mark and tick off every 2″ (Fig. 27)?

FIG. 27

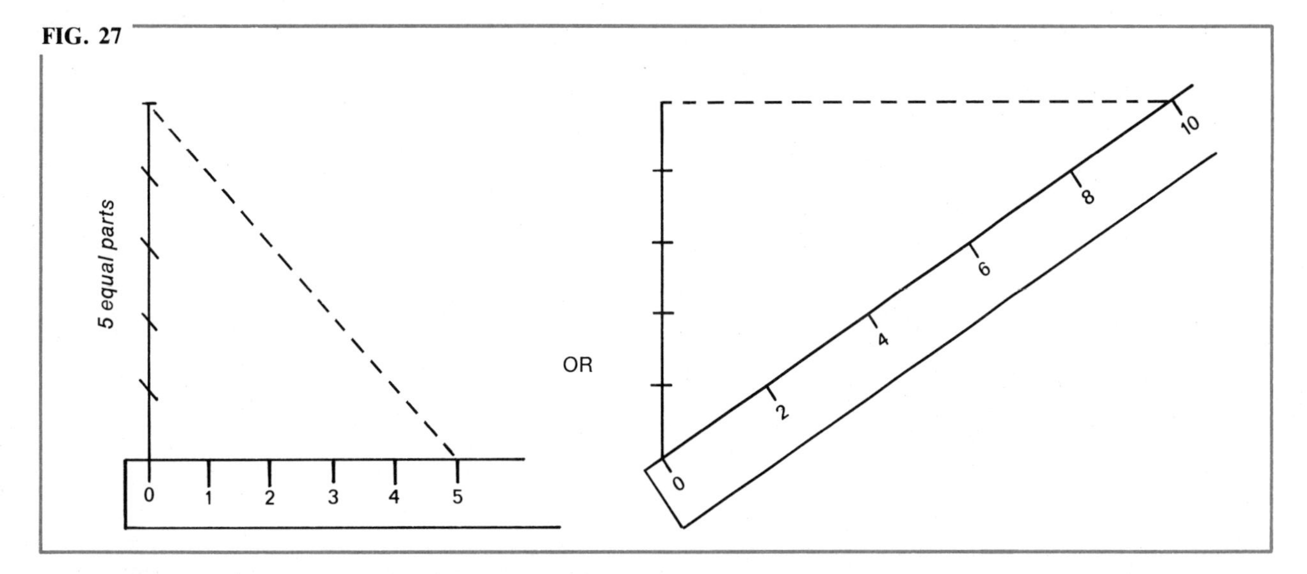

This is the geometric principle of constructing a scale.

After all is said and done, what is important is how you use your ruler rather than which ruler you use. Forget the original purpose of a ruler, which is to measure inches and parts of inches. It shouldn't take long to adjust to seeing it as a stick with marks on it. Once you understand this, you can use other types of rulers in the same way. As a matter of fact, there are other rulers better suited for this purpose. The engineer's ruler, for instance, is probably the most useful of all. Next would be a pica ruler, agate ruler, or centimeter ruler.

Each of these rulers has been designed for a specific purpose. While they may have the same overall length, each offers a different set of segments. However, they all represent a line broken into some number of equal divisions. Study them for a while. You may already be familiar with some, if not all of them. Figure 28 shows a section of each, all of the same length.

FIG. 28

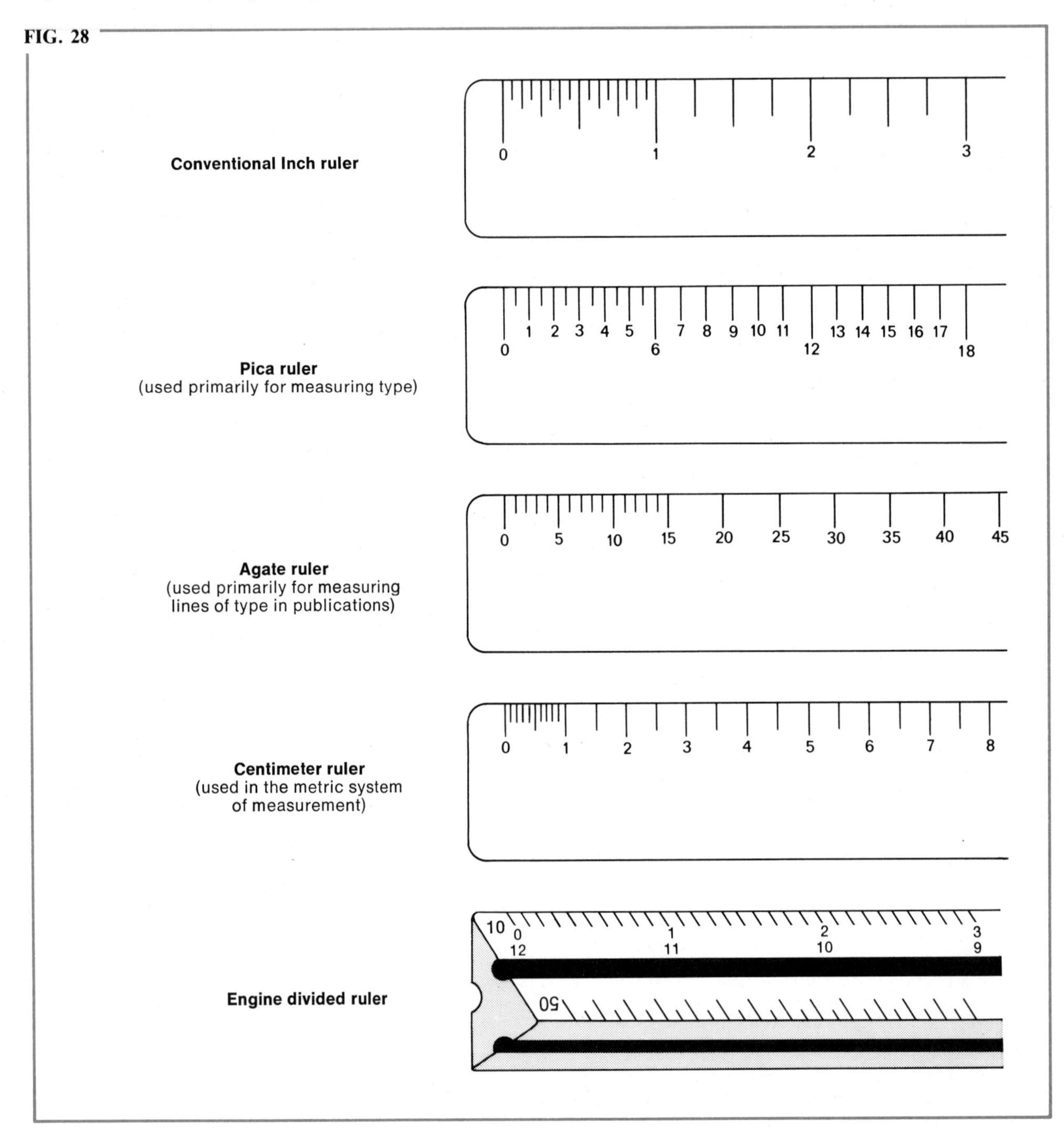

To understand better how you can use these rulers, let's construct the same scale with each of them (Fig. 29).

FIG. 29

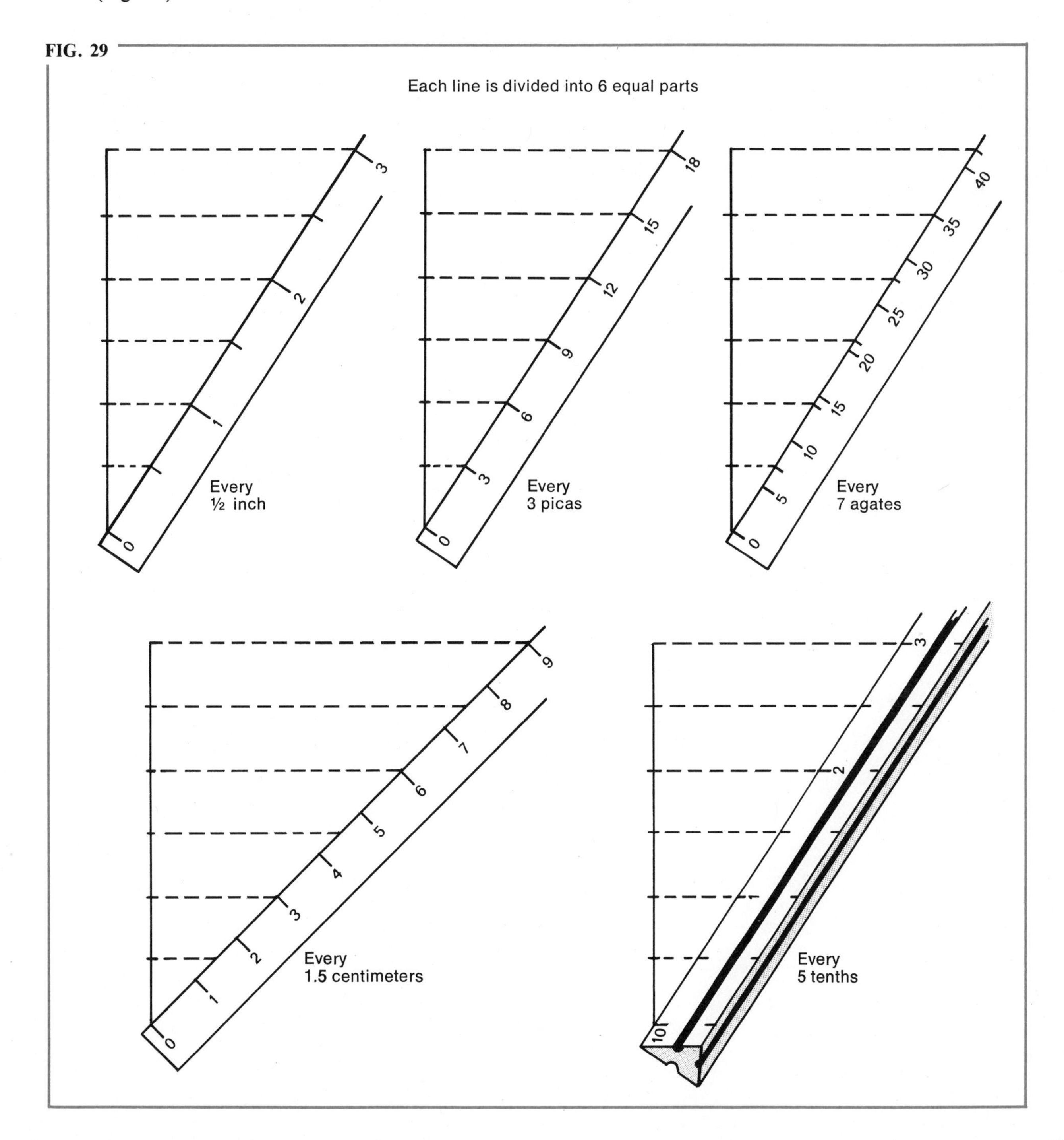

In the final analysis it should be obvious that any ruler can be used. It is all a matter of preference based on the requirements of the chart. But once you have gotten deeper into this subject, you will almost certainly use the engineer's ruler more often than any other. As a matter of fact, for the sake of simplicity, most of the exercises and examples in this text will be done with the engineer's ruler. Also, it makes absolutely no difference whether you measure in inches or in metric units. The principles are the same. But, occasionally throughout the book, a comparison will be demonstrated.

Primarily you will be working with relatively simple geometric formulations in conjuction with basic arithmetic.

It is strongly recommended that you do not start the next chapter until you have mastered this one. If you do, you will only become confused, think it is all impossible, and give up. If you take it step by step, one chapter at a time, not only will you learn a fascinating skill, but you will also be in a better position to earn better pay, and I will have fulfilled my purpose in writing this book.

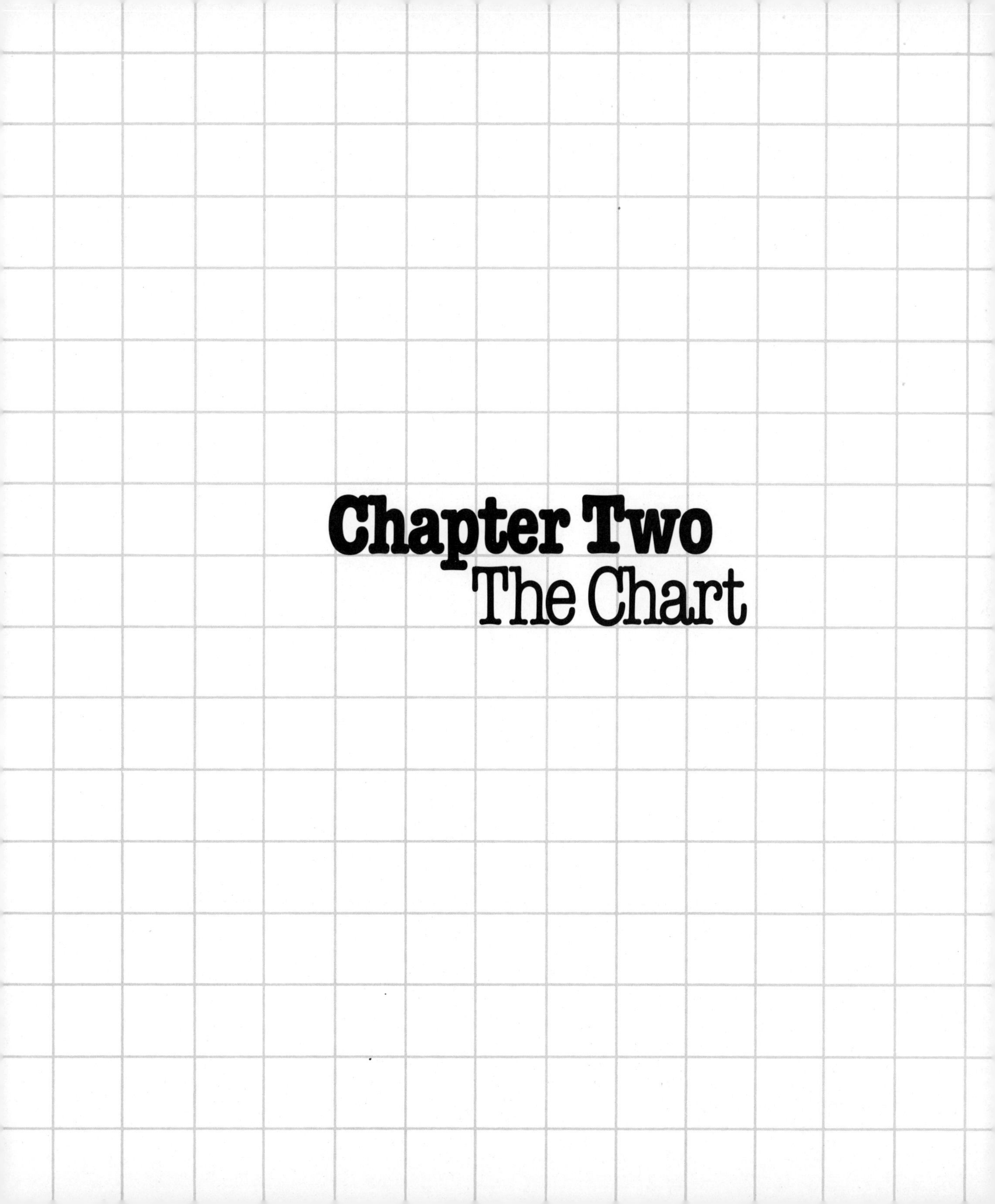

Chapter Two
The Chart

Now that you have been introduced to the ruler and how to construct a scale, you are better equipped to learn the next phase. However, you are not quite ready to draw a chart at this time. There are a few more principles and formulas you need to become familiar with. Above all, be patient and follow these steps very carefully. Do not rush through them.

While studying these methods and doing the exercises, it may be best to use ledger bond paper rather than expensive drawing paper or illustration board. Whatever you draw on, be certain it is taped or tacked firmly on your drawing table.

SPACE DIVISION

Using your T-square, triangle, a good ruler, and a 6H pencil, construct a rectangle *exactly* 4″ wide by 2″ high. Be sure you allow enough space around your drawing. Construct it just below the center of your paper. The paper for these exercises could be 9″ by 12″, but you will find 11″ by 14″ paper easier to manage. (Fig. 30).

FIG. 30

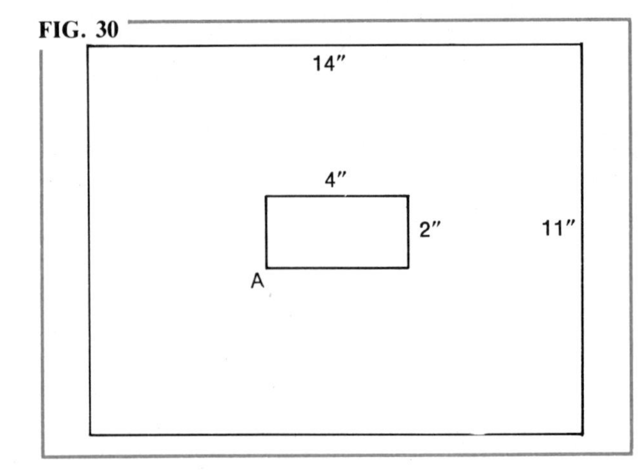

The problem: Divide the rectangle into 3 equal parts on the width and 3 equal parts on the height.

Place your ruler on the base line, the 0 mark at A. You need to find 3 equal segments on your ruler. The 3″ mark falls short of the width, so you will have to use the 6″ mark and count off every 2″. The 6″ mark extends beyond the width. So pivot your ruler (0 at A) until the 6″ mark touches the right vertical. You will notice that it is still too long to touch the line (Fig. 31).

FIG. 31

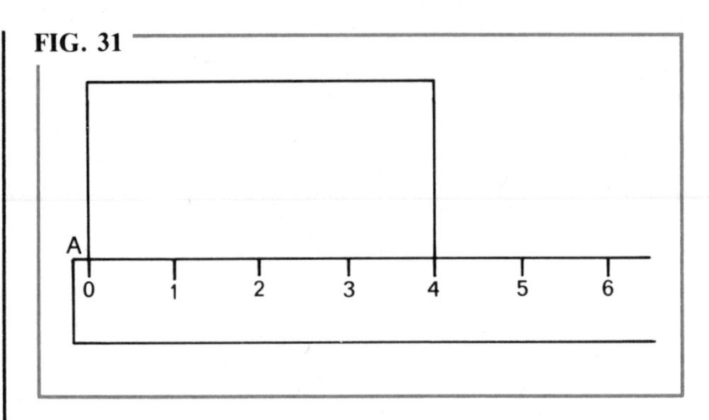

Whenever this occurs, extend the right vertical line a few inches beyond the 2″ mark (Fig. 32).

FIG. 32

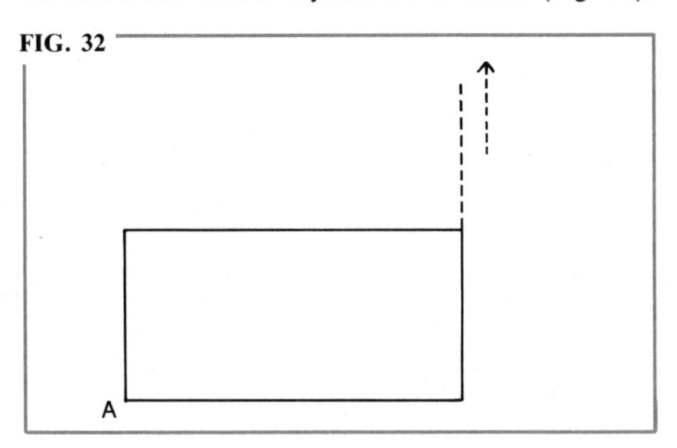

Reposition your ruler and pivot it until the 6″ mark touches the extended vertical line (Fig. 33).

FIG. 33

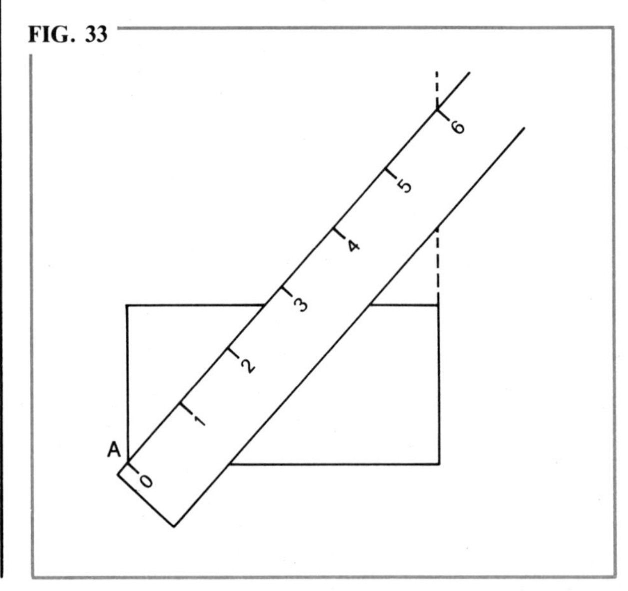

Hold the ruler firmly in this position while you draw a light pencil line along the edge of the ruler. Do not move the ruler! Now, *lightly* tick off every other inch (2, 4, 6) (Fig. 34).

FIG. 34

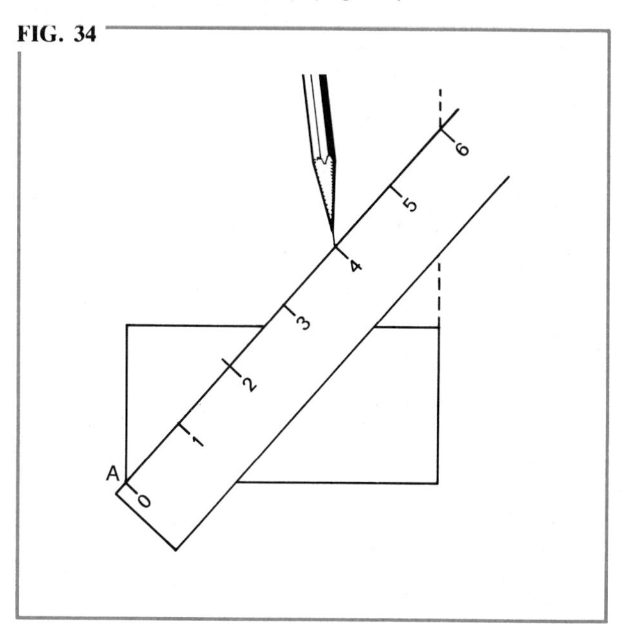

Remove the ruler. Set up your T-square and triangle. Align the triangle with the first tick mark and draw a vertical line through it. Repeat this with the second tick mark (Fig. 35).

You have divided the width into 3 equal parts. Now to divide the height into 3 equal parts.

The 3″ mark would be the first candidate, but it falls beyond the top of the rectangle. Keep the 0 mark at A and pivot the ruler until the 3″ mark touches the top horizontal line. Draw a light line along the edge of the ruler and tick off the 1″ and 2″ marks (Fig. 36). Remove the ruler, set up your T-square, and draw a light horizontal line through each tick mark. You have divided the height into 3 equal parts (Fig. 37).

Let's stay with this problem a moment longer. Construct another 4″ by 2″ rectangle on a fresh sheet of paper. Suppose you were to divide the height into 14 equal parts. With the 0 mark at A, the 7″ mark could be used (counting off every $\frac{1}{2}$″ mark), but it extends beyond the rectangle. Extend the top *horizontal* line and pivot your ruler until the 7″ mark touches it (Fig. 38). Then tick off every $\frac{1}{2}$″ mark.

FIG. 35

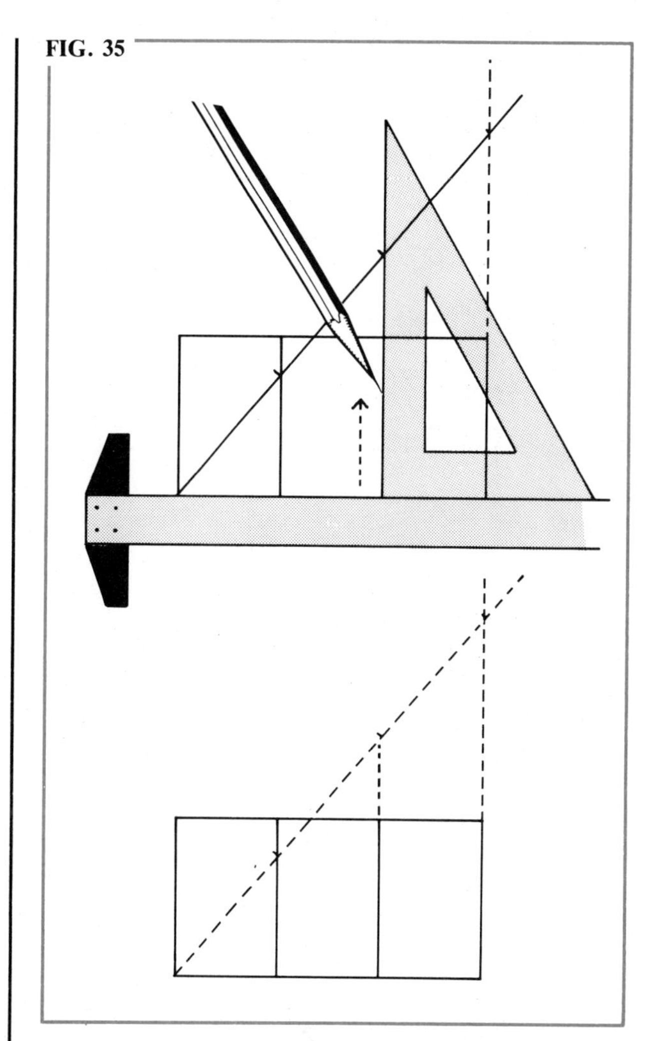

FIG. 36

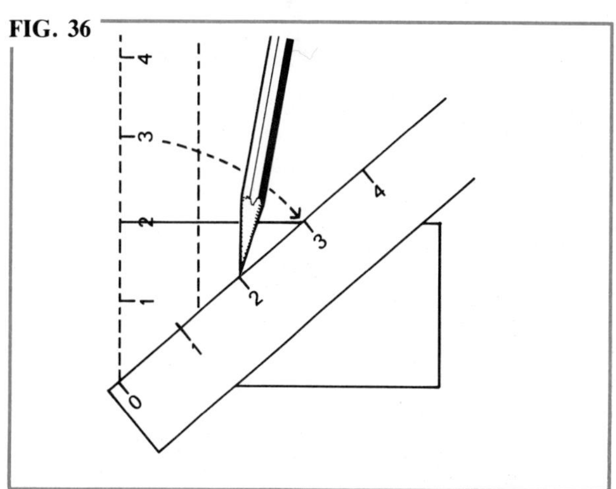

FIG. 37

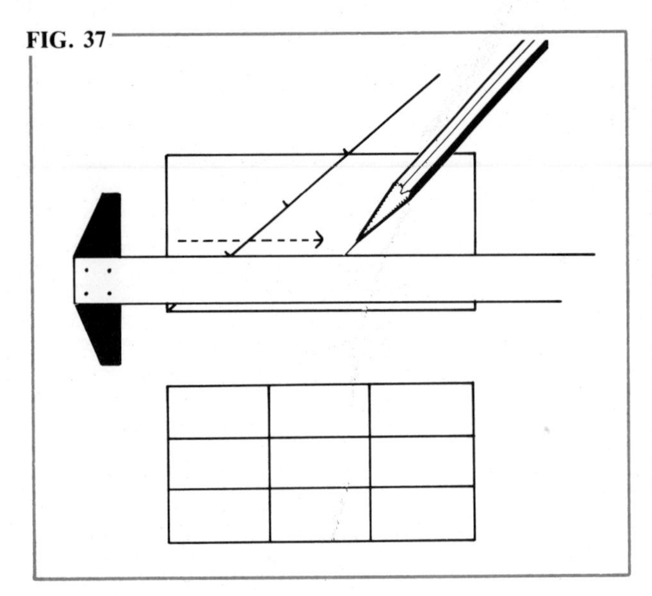

FIG. 38

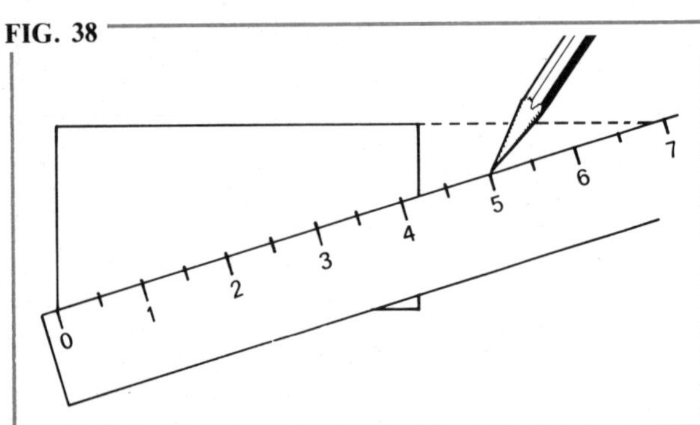

Draw a horizontal line through each tick mark (Fig. 39).

FIG. 39

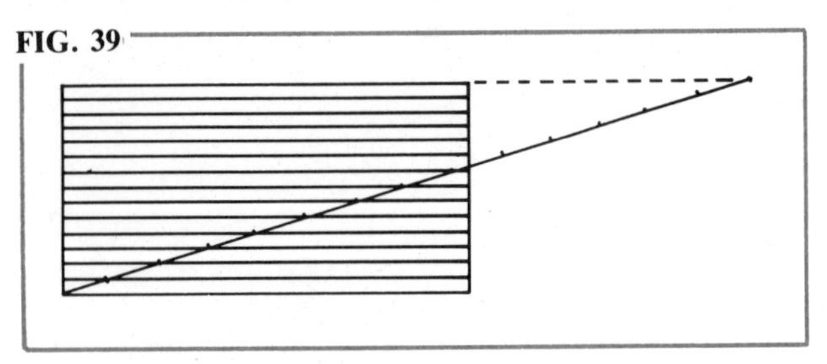

Try another. You must continue to practice this principle until it becomes second nature.

Construct a $5\frac{3}{8}''$ by $4\frac{1}{16}''$ rectangle.

Problem: divide the $4\frac{1}{16}''$ height into 11 equal parts and the $5\frac{3}{8}''$ width into 26 equal parts. Once again, place your 0 mark at A.

Pivot the ruler until the $5\frac{1}{2}''$ mark touches the *top horizontal* line. Draw a light line and tick off every $\frac{1}{2}''$ mark (Fig. 40).

FIG. 40

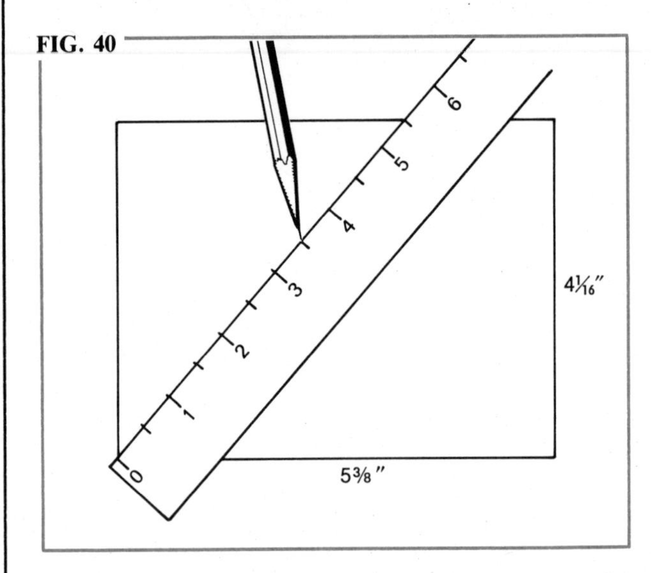

Remove the ruler, set up your T-square, and draw a light horizontal line through each tick mark. The result will be 11 equal divisions on the height (Fig. 41).

FIG. 41

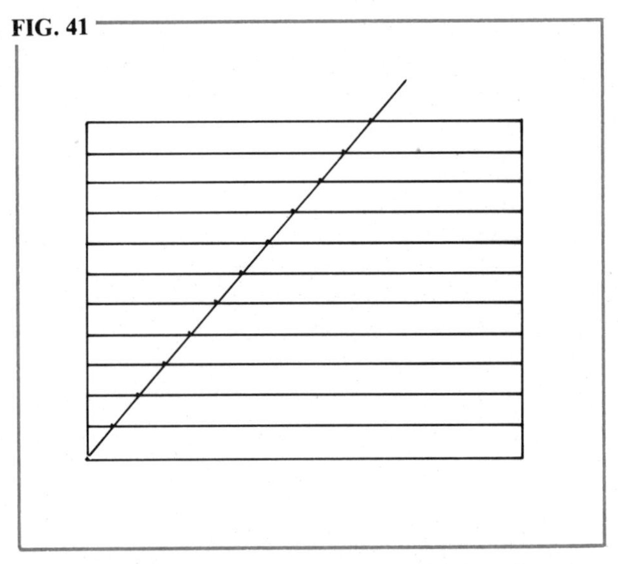

Now for the 26 parts on the width. Align the 0 mark at A and pivot your ruler until the $6\frac{1}{2}''$ mark touches the *right vertical*. Draw a light line and tick off each $\frac{1}{4}''$ mark (Fig. 42).

FIG. 42

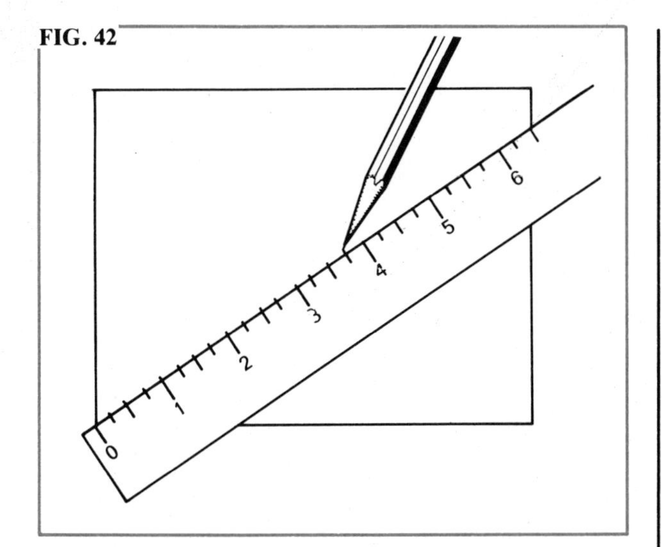

Remove the ruler, set up your T-square and triangle, and draw a vertical line through each tick mark. The result will be 26 equal divisions on the width (Fig. 43).

FIG. 43

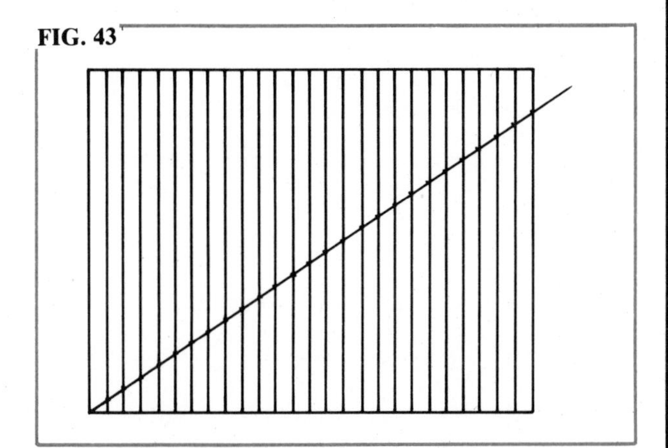

By now you should be developing an understanding of the principle of geometric construction. Note that in each problem you are working with a diagonal line. What you are doing is projecting measurements from a strategically positioned diagonal line. Not surprisingly, many artists refer to this method as the *diagonal line method.*

Using the diagonal line this way is, in itself, the main principle employed in the approach to chart making. You will need to acclimate yourself to the logic of this somewhat abstract approach. After several exercises it should begin to seem very simple. As a matter of fact, it is so logical you may even forget there was a time you didn't know it at all. Nevertheless, you will need considerable practice and intense concentration to master it.

Here is another problem.

Construct a rectangle 3″ by 7″.

The problem: divide 3″ into 10 equal vertical parts and the 7″ into 5 equal horizontal parts (Fig. 44).

FIG. 44

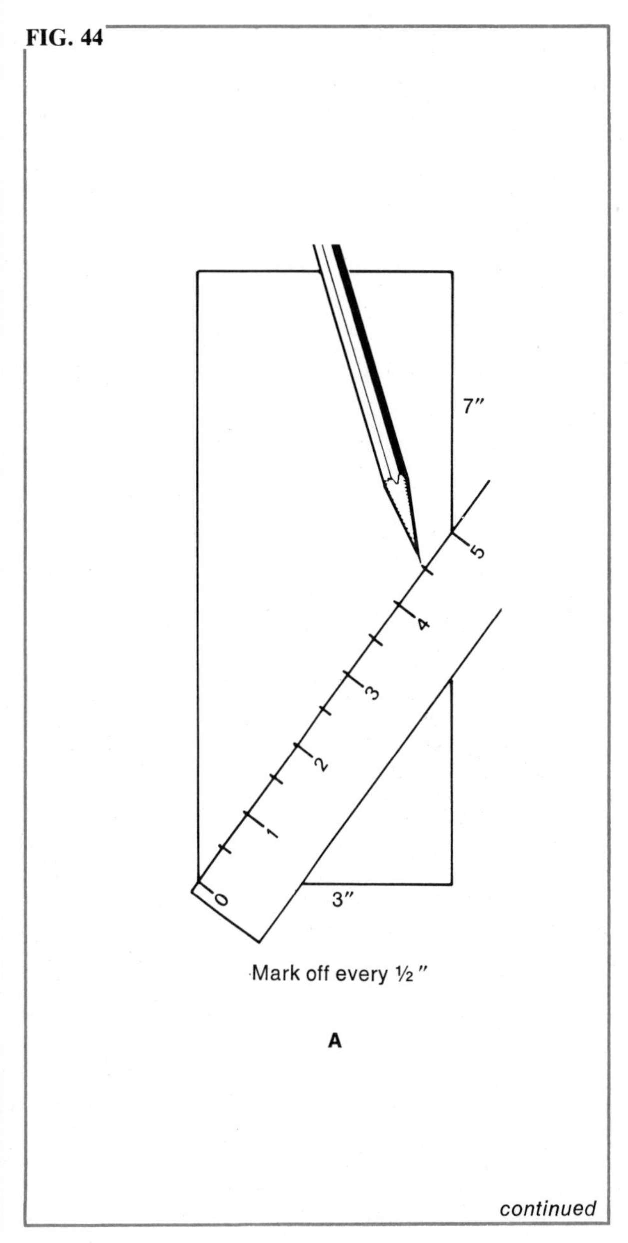

continued

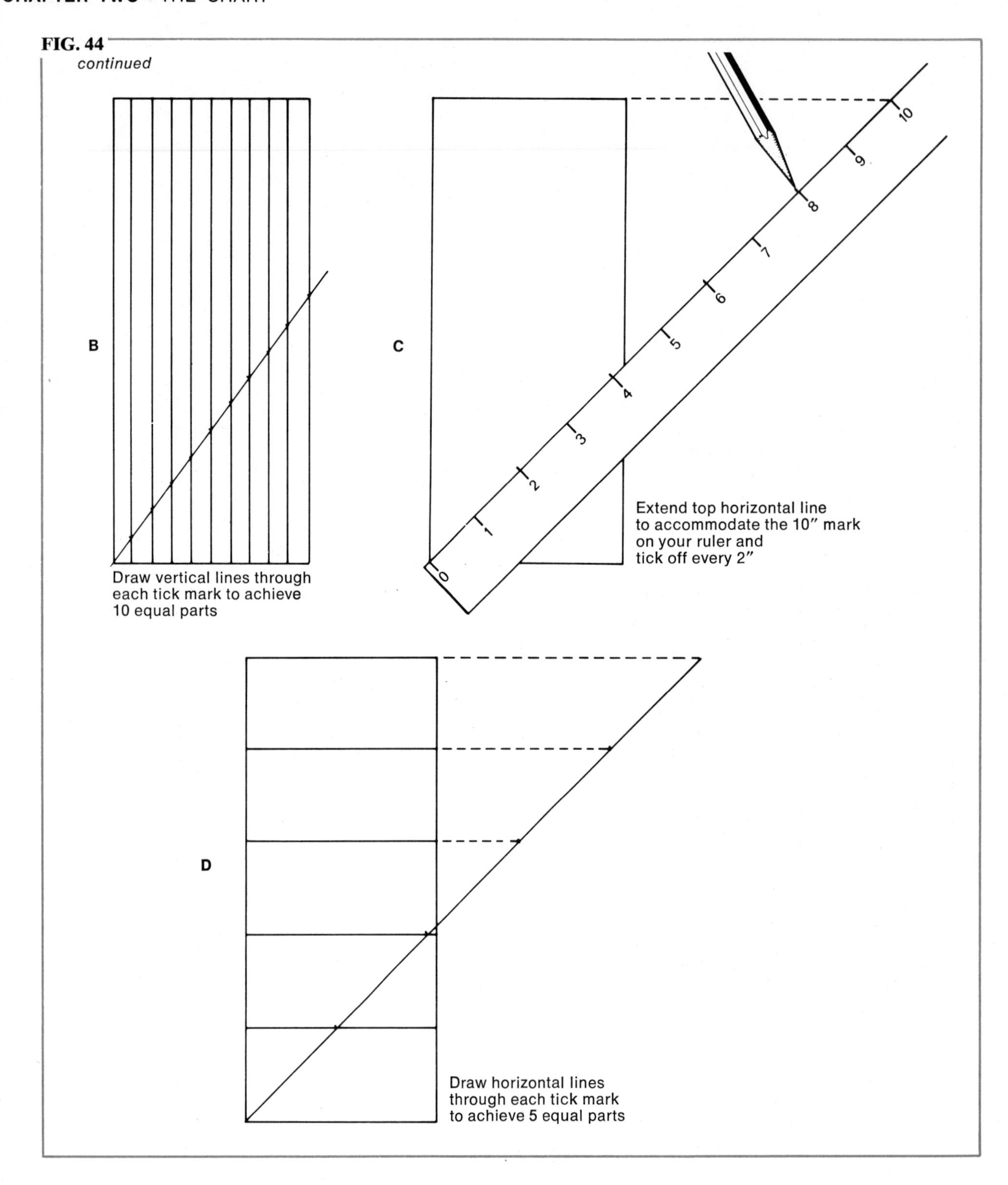
FIG. 44
continued
B
Draw vertical lines through each tick mark to achieve 10 equal parts
C
0
1
2
3
4
5
6
7
8
9
10
Extend top horizontal line to accommodate the 10″ mark on your ruler and tick off every 2″
D
Draw horizontal lines through each tick mark to achieve 5 equal parts

Now for one that requires a little more concentration.

Construct a 7″ by 1″ rectangle.

The problem: divide the 1″ height into 12 equal parts (Fig. 45).

FIG. 45

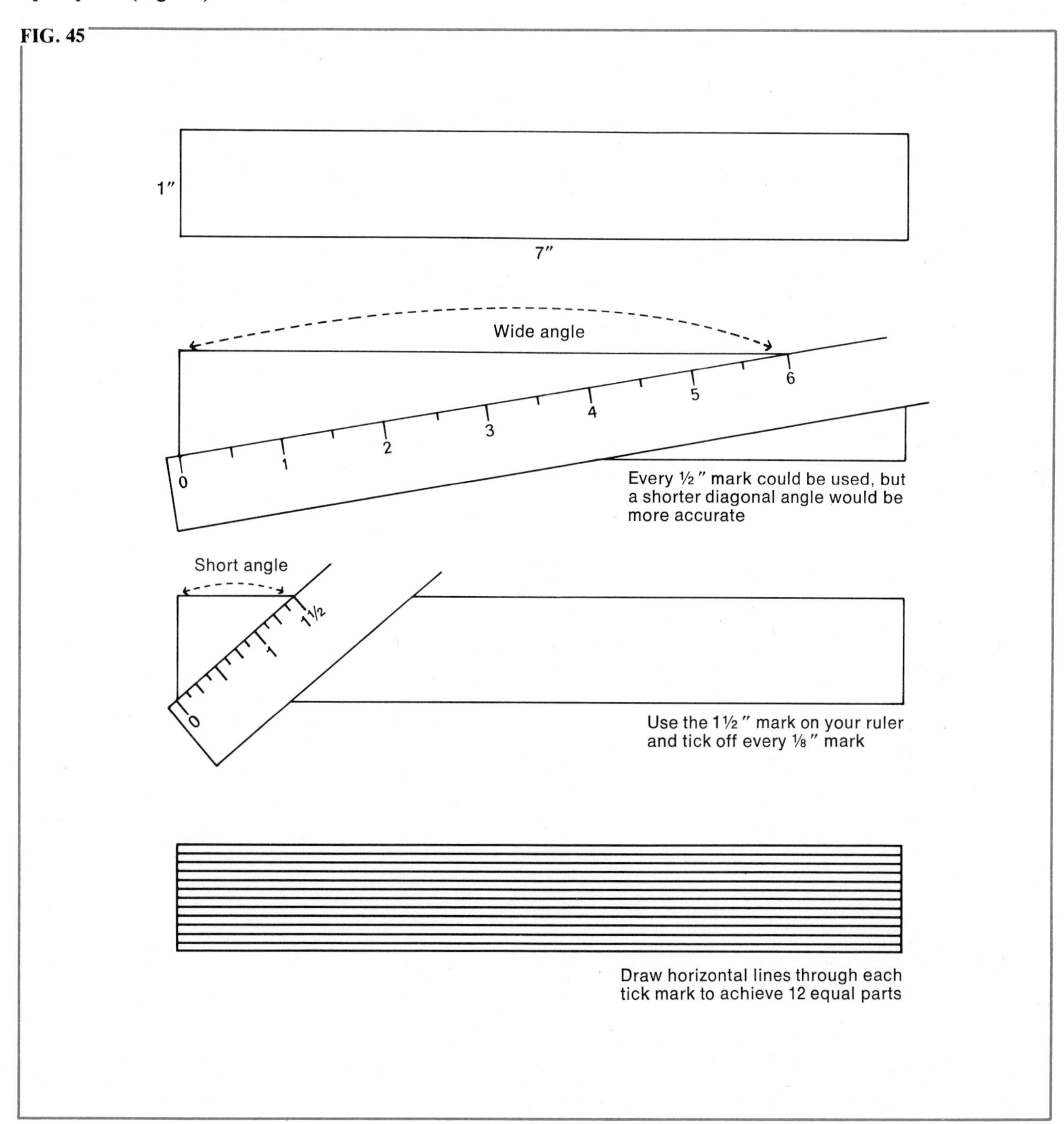

This one requires a little extra thinking. Using the same size rectangle, divide the 1″ height into 14 parts. But they will not all be equal. The eighth part is to be exactly half of the other 13 equal parts (Fig. 46). Test yourself at this time. Do not look at the solution until you've tried it first. If you are wrong, it is advisable to go back over the last two or three problems before continuing. If you are correct, you are on your way.

FIG. 46

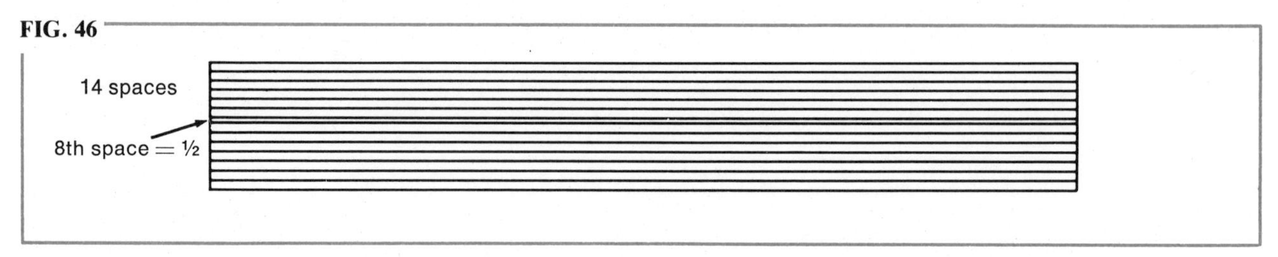

Actually, you need a total of $13\frac{1}{2}$ spaces, which can be found quickly on your ruler at $13\frac{1}{2}$″. You will need a lot of room to do it using $13\frac{1}{2}$″, but try it anyway. Extend the right vertical line until you can line up the $13\frac{1}{2}$″ mark.

Position your ruler 0 mark at A. Pivot until $13\frac{1}{2}$″ touches the top (extended) horizontal line (Fig. 47).

FIG. 47

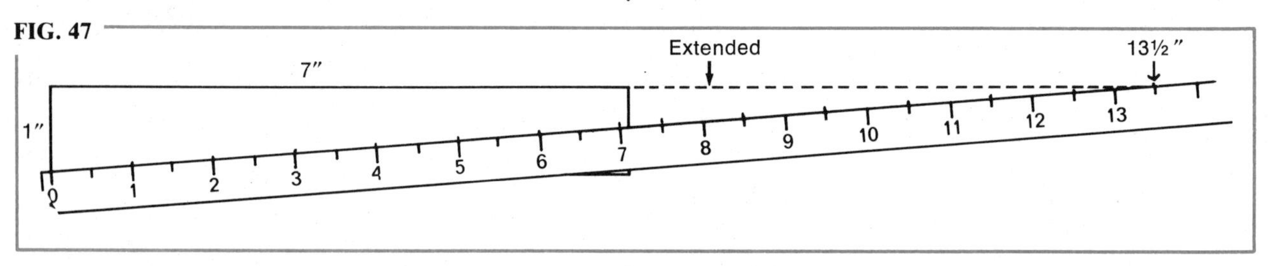

This is where you must think.

Tick off the first seven 1″ marks. Then tick off the next $\frac{1}{2}$″ mark (at $7\frac{1}{2}$″). From that point on, tick off every inch on the $\frac{1}{2}$″ mark ($8\frac{1}{2}$″, $9\frac{1}{2}$″, etc.) to $13\frac{1}{2}$″. Remove your ruler and draw a horizontal line through the tick marks. Note that the eighth division is equivalent to half a division (Fig. 48A).

Perhaps there is another segment you could use rather than $13\frac{1}{2}$″, which is quite a wide diagonal. Or maybe you only have a 12″ ruler. All right, try $\frac{1}{2}$″ marks. $13'' \times \frac{1}{2} = 6\frac{1}{2}''$, to which we add $\frac{1}{2} \times \frac{1}{2}'' = \frac{1}{4}''$, for a total of $6\frac{3}{4}$″. Use the $6\frac{3}{4}$″ mark on your ruler and mark off the first seven $\frac{1}{2}$″ marks, then the next $\frac{1}{4}$″, and from there on every $\frac{1}{2}$″. The eighth division will automatically be half of each of the others (Fig. 48B).

FIG. 48

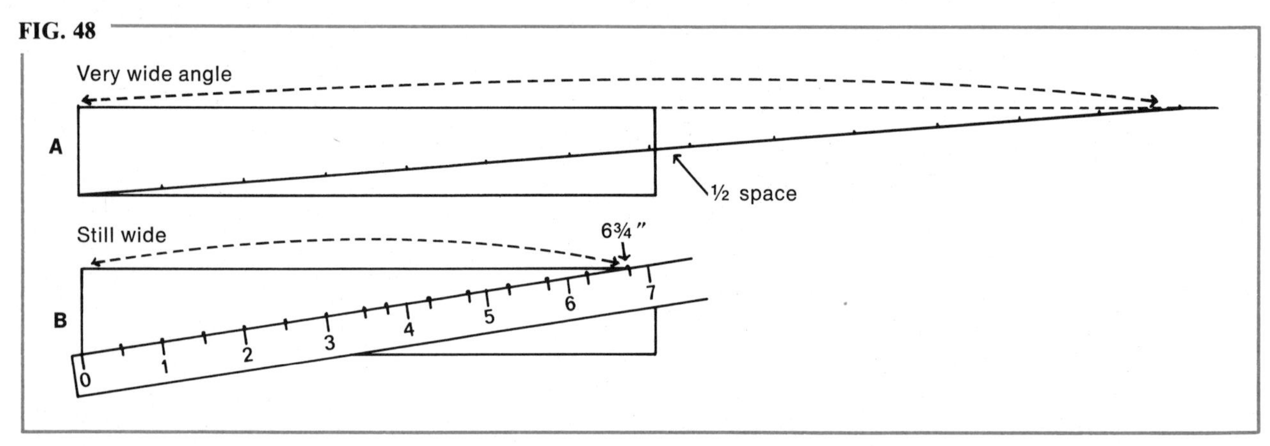

The angles for both the $13\frac{1}{2}''$ and $6\frac{3}{4}''$ are still quite wide and not as accurate as they need to be.

Remember, always try for the least shallow angles. How about $1\frac{11}{16}''$ (Fig. 49)?

FIG. 49

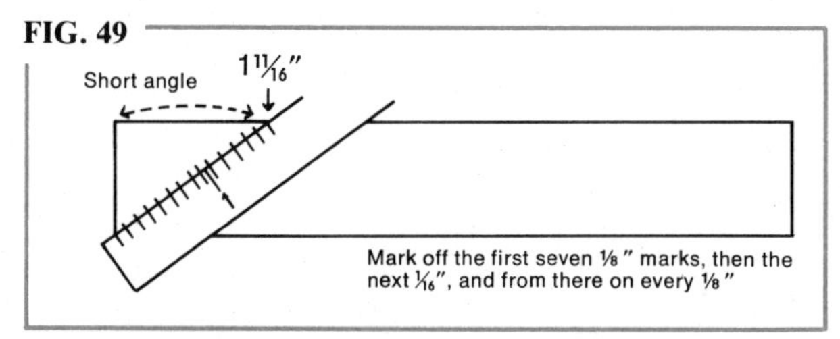

You should be ready now for something more advanced.

Construct a rectangle $7\frac{7}{16}''$ by $5''$. (Note that, as a general rule, when giving the dimensions of a rectangle the first number should be the width, the second the height.)

Problem: divide the rectangle in half on the width.

Here are four approaches to this relatively simple problem:
(Fig. 50)

A. Divide $7\frac{7}{16}''$ in half mathematically.
B. Draw a diagonal line from each corner to the opposite corner. Then draw a perpendicular line from the base line through the intersection of the diagonal lines.
C. Take any point on your ruler that is easily divided in half. In this problem, 8″ could be used. Align the 0 mark on the left vertical and pivot the ruler until the 8″ mark touches the right vertical. Tick off the 4″ mark. Draw a perpendicular line from the base line through the tick mark. This is a very quick and useful method when working with odd sizes.
D. This next method could easily be referred to as a trick of the trade. Aside from being accurate, it is exceptionally useful when dividing small, odd-sized lengths in half.

Take any sheet of paper that has a straight edge. Place the straight edge of the paper on the base line so that one end of it touches the beginning of the line. Tick off the other end of the line with your pencil. Now fold the paper so that the corner touches the tick mark. Press the fold to crease the paper. Open it. The crease is the center. Line up the paper with the pencil line, tick off the center mark and draw a perpendicular line up from that point.

FIG. 50

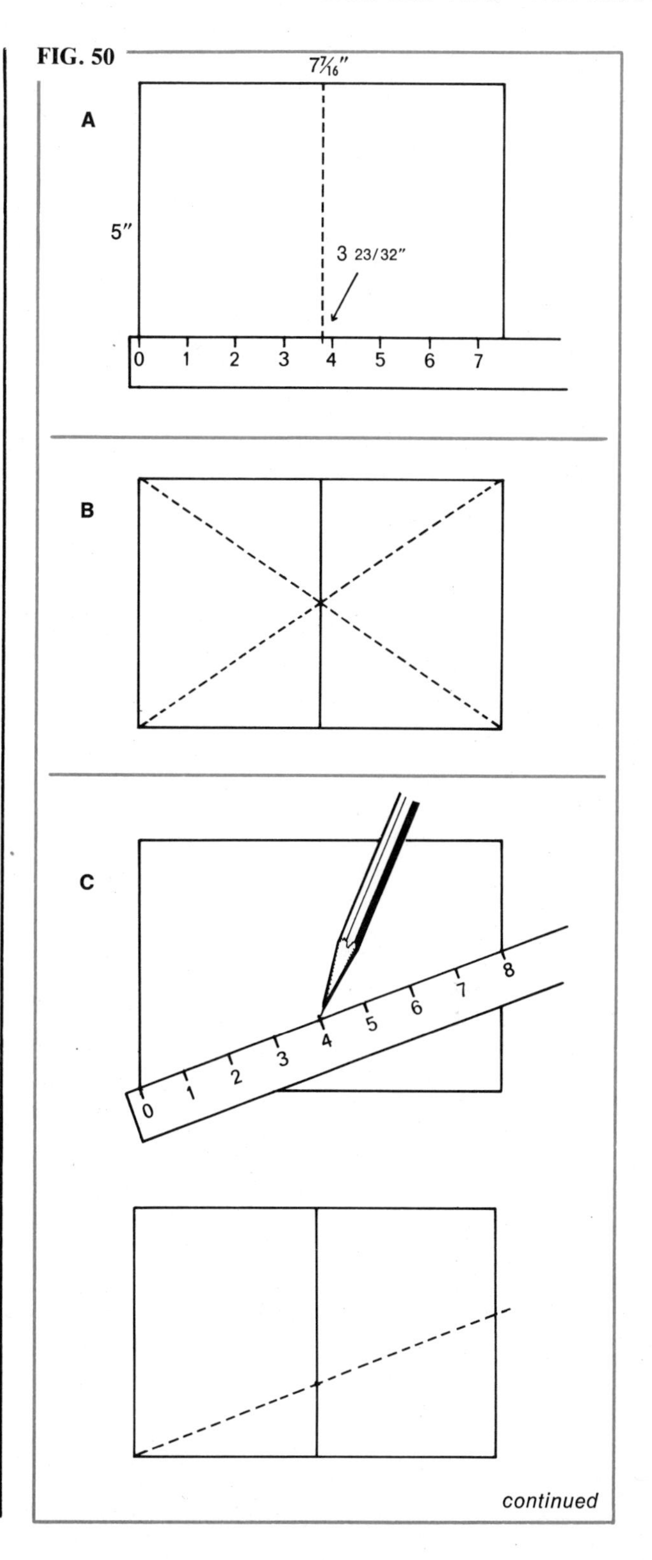

continued

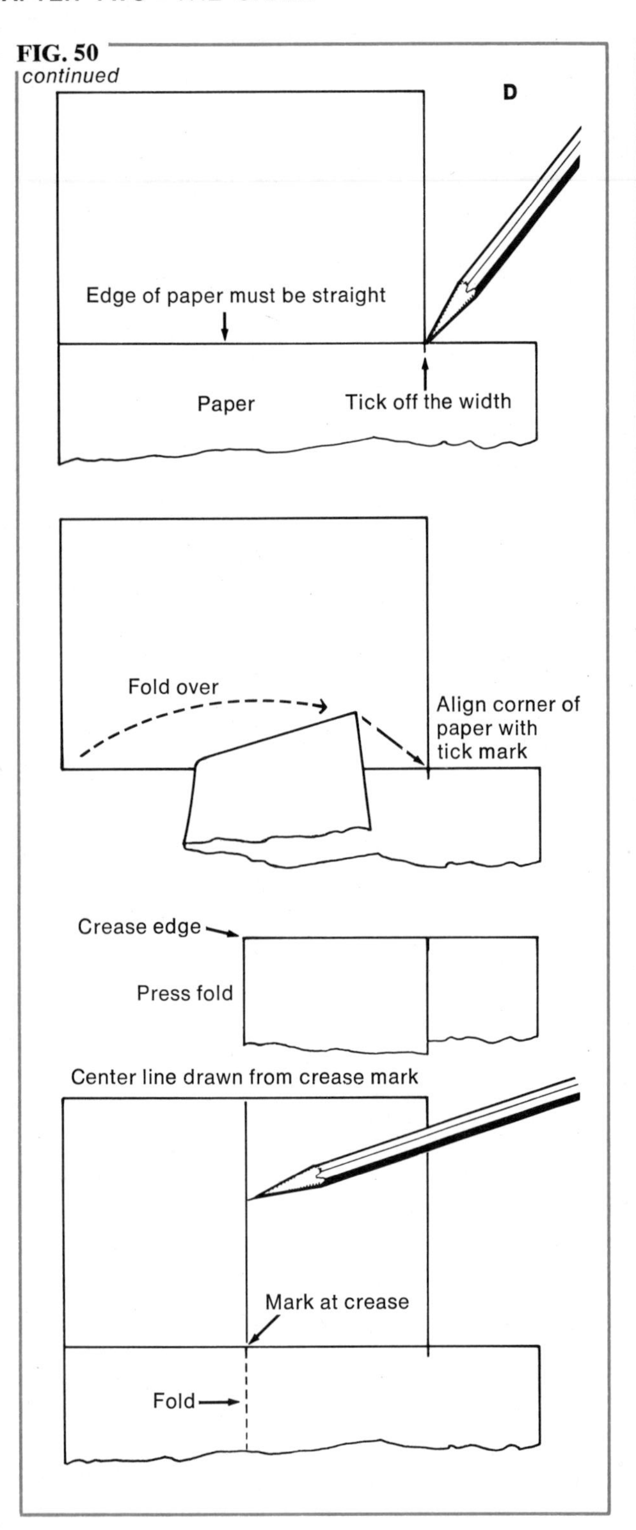

FIG. 50 *continued*

Of course, there are other methods that could be used to divide a given line in half, but those just described seem to be the most appropriate.

Now that you have divided the width in half, divide the left half into 4 equal parts on the width, and the right half into 5 equal parts on the width (Fig. 51).

FIG. 51

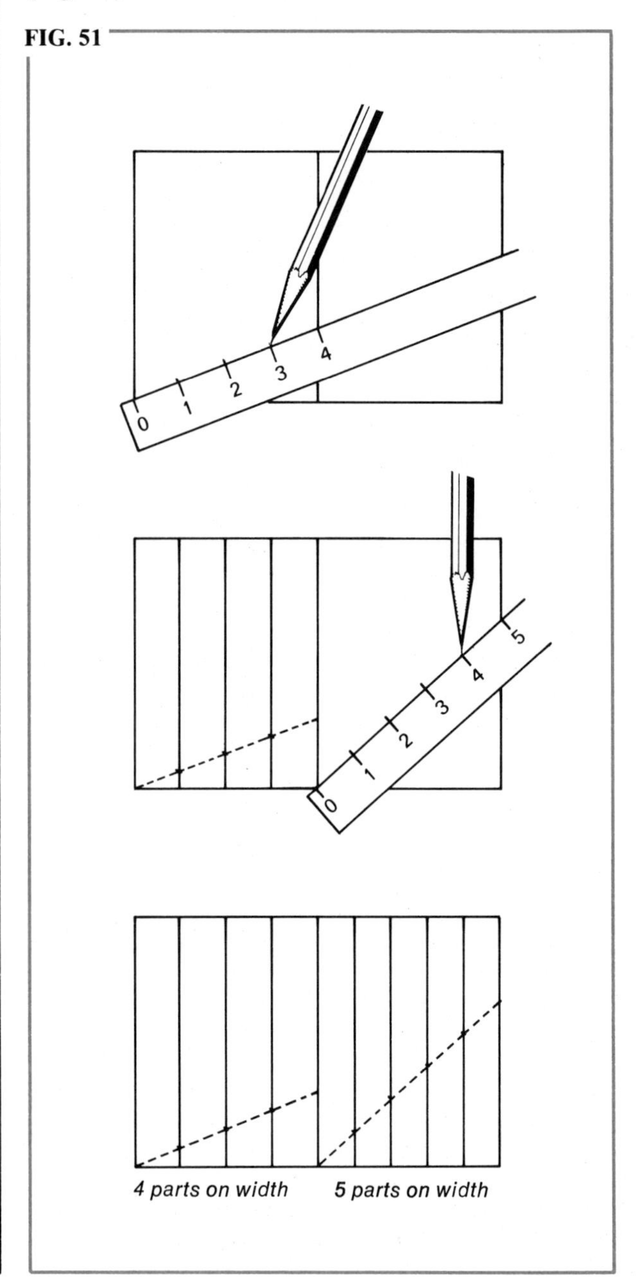

Next, divide the *height* into 11 equal parts (Fig. 52).

FIG. 52

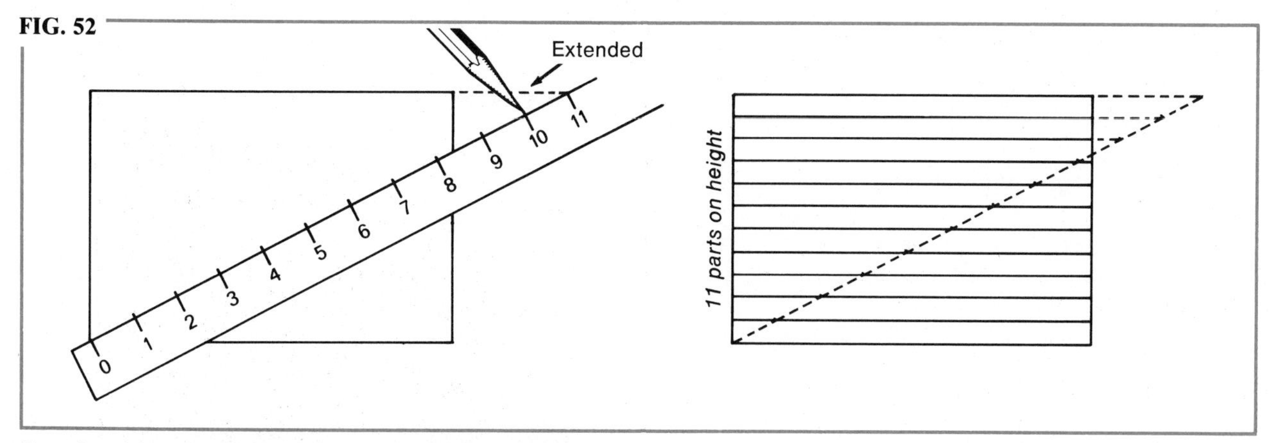

Try the same exercise with your centimeter ruler (Fig. 53).

FIG. 53

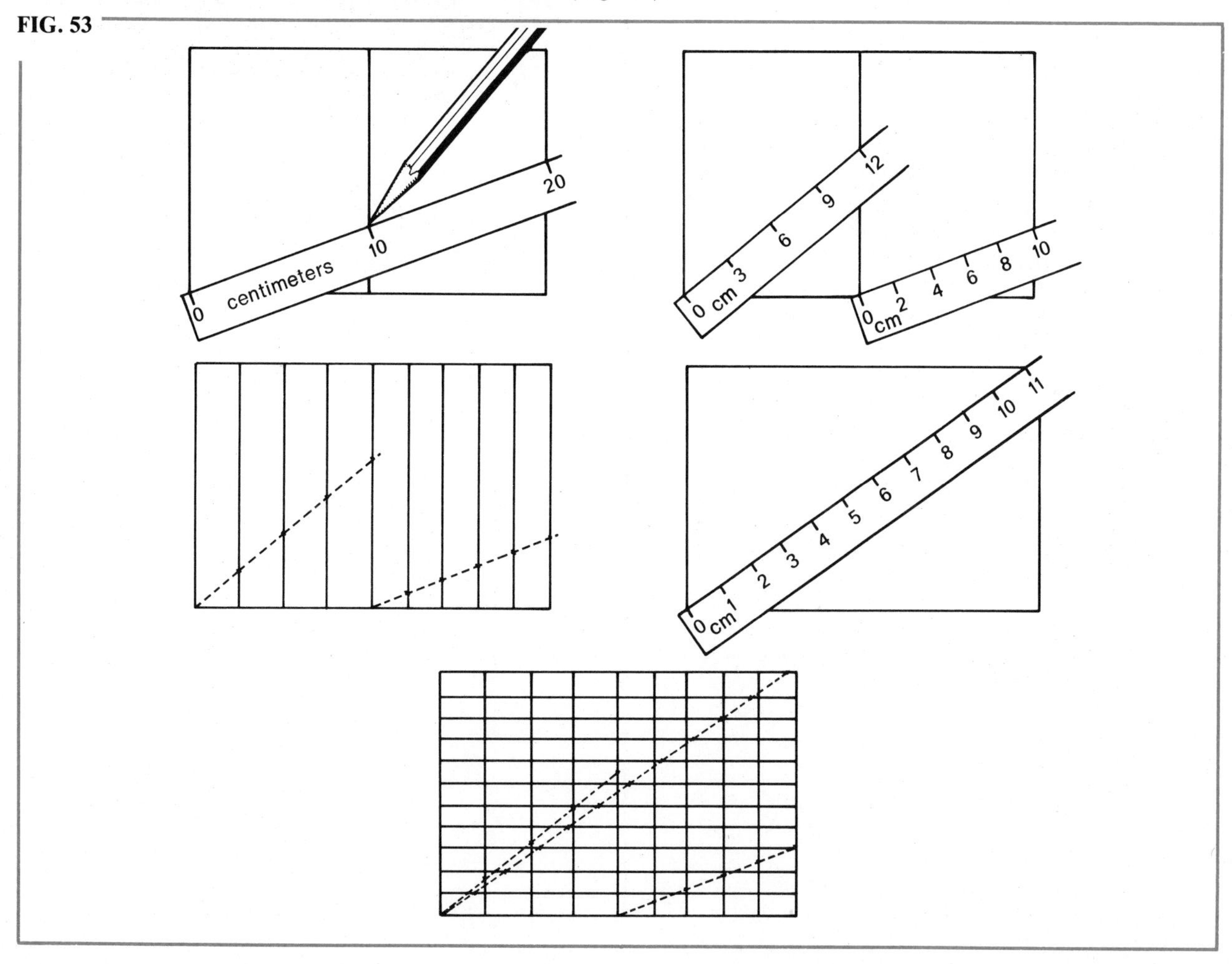

Now you are ready to construct your first chart. It will be a simple one because there is still one more very important thing to learn. You will see when we get to it.

The specifications are as follows:

A population comparison for the years 1965, 1970, and 1975.

The figures are 1965: 140,000; 1970: 172,000; and 1975: 188,000. It is to be a column chart.

Before you start a chart, be certain you have all the information you need and analyze it as thoroughly as possible.

Your first consideration might be how high (in thousands) the scale should be. Also, how many increments should the scale contain? For the most part the top figure on the scale is determined by the tallest bar, in this case 188,000.

The highest point on this chart could be the top of the tallest bar or a slightly larger, rounded number such as 190,000 or 200,000. The scale is to be divided into a specific number of increments. This is usually determined by the number of bars, the figures each bar will represent, or the span between the lowest and highest figures.

The next consideration is the width. You will need to know the overall dimensions (width and height) of the chart. Often you will have a given area to work within. At times the dimensions will be specific, or you may have the freedom to prepare it as you see fit.

For obvious reasons the exercises in this text will be specific. The overall dimensions for this chart are $3\frac{1}{8}''$ by $3\frac{5}{8}''$. At this time you should make a rough sketch of what the chart should look like based on all the information you have, regardless of how simple it may be (Fig. 54).

Let's box the chart off at a 200,000 height. Increments of 25,000 would be small enough for this chart.

The first and last column will be $\frac{1}{4}''$ in from the outside edges as well as $\frac{1}{4}''$ between each column.

Use a fresh piece of ledger bond (9″ by 12″ or 11″ by 14″) for this exercise, positioning the chart visually in the center.

The following step-by-step procedure is one that should be memorized. It is relatively simple and applicable to most charts.

Before you start, gather all the tools you need: a T-square, a 10″ triangle (30° 60° 90°), a 6H pencil, a ruler (no shorter than 12″), an engine

FIG. 54

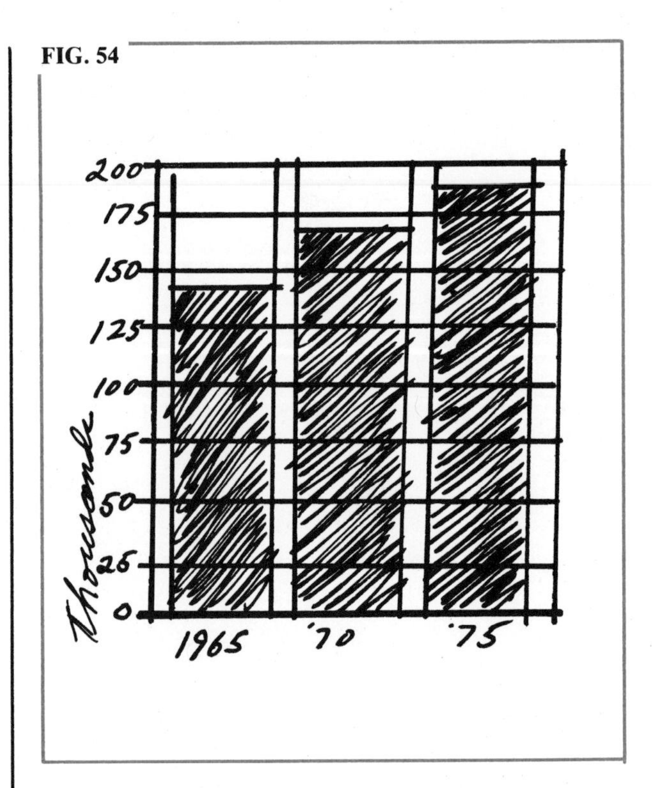

FIG. 55

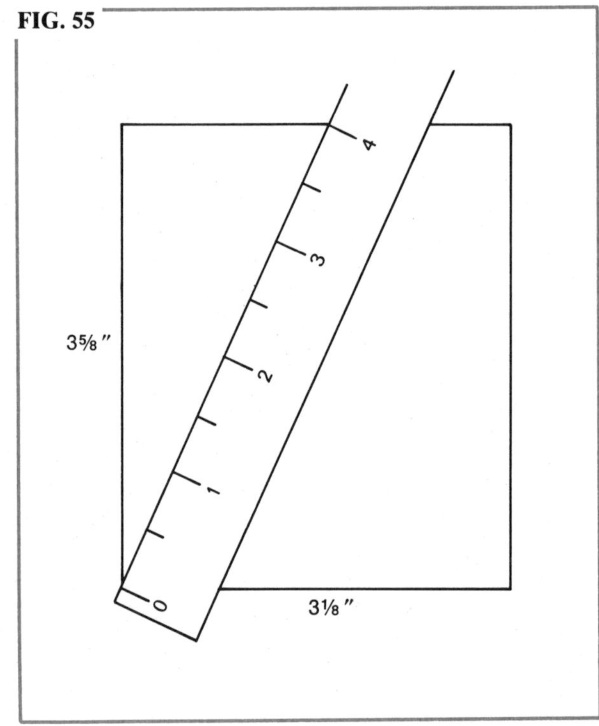

FIG. 56

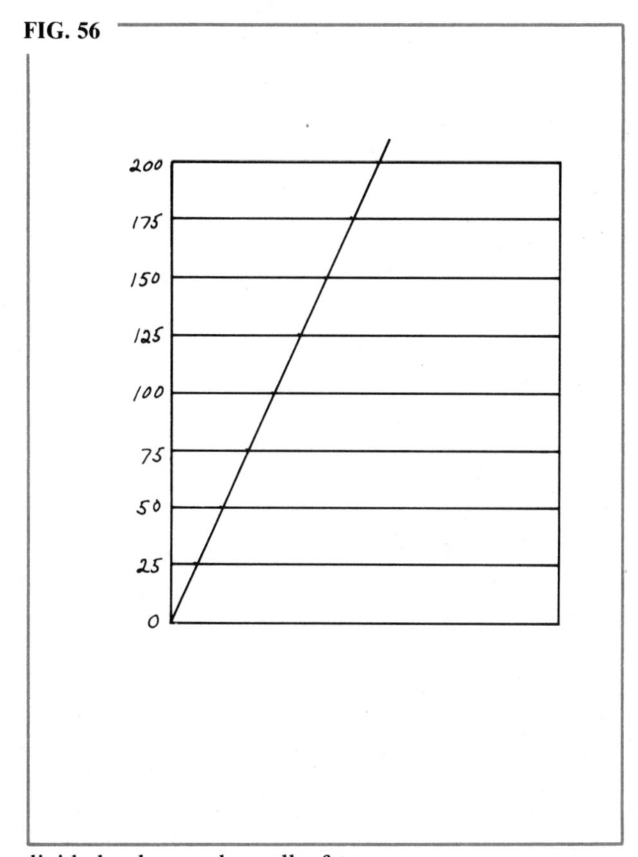

divided ruler, and a roll of tape.

First construct the overall dimensions. Now create the scale along the left side. It will be divided into 8 equal parts. Use your regular ruler if you choose to and position the 0 mark at the lower left corner of the chart. The 4″ mark can be used, but the height of the chart is too short. So pivot the ruler until the 4″ mark touches the top horizontal. Draw a light line along the edge of the ruler and tick off every $\frac{1}{2}''$ mark for a total of 8 spaces (Fig. 55). Please note: When working with charts, the word *space* is common terminology. During the process of constructing a chart the maze of tick marks and guidelines is often confusing. Therefore, the word space is used to distinguish the distance between lines, bars, columns, etc., as opposed to the width of a bar, column, etc. Example: space, column, space.

Set up your T-square and draw a light horizontal line through each tick mark (Fig. 56).

Lightly label each line in pencil so that it will be easier to identify them later on. What you have

FIG. 57

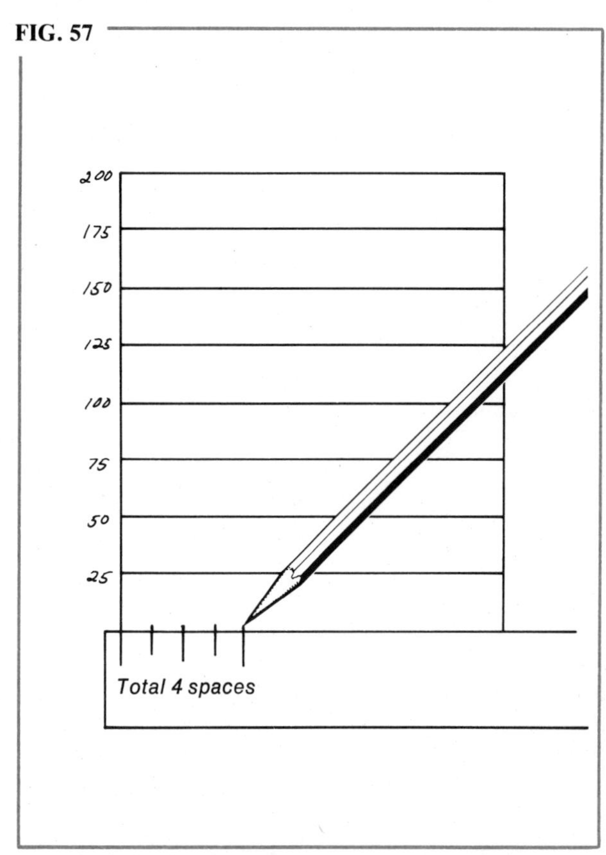

done here is simultaneously create two things: the scale along the left vertical and a *grid* across the entire chart area. This grid is actually an extension of the scale to be used in calculating the heights of the columns.

Now to establish the positioning of the columns. What you are about to do requires a bit of concentration, so follow it carefully. You need a total of 3 columns and a total of 4 spaces (the length of the space between columns and the outside edges). If the spaces are established to be $\frac{1}{4}''$, the problem is: how wide should the columns be?

Position your ruler 0 mark at the beginning of the base line and tick off 1″. This mark represents the total of the four $\frac{1}{4}''$ spaces. You have now subtracted what represents all the spaces from the overall length of the line (Fig. 57).

Now that you have subtracted all the spaces, the remaining portion of the line represents the total for the three column widths. Now, position your ruler 0 mark at the 1″ mark and pivot until the 3″ mark touches the right vertical. All that is necessary is to tick off one increment, because you only need to know the width of one column (Fig. 58).

FIG. 58

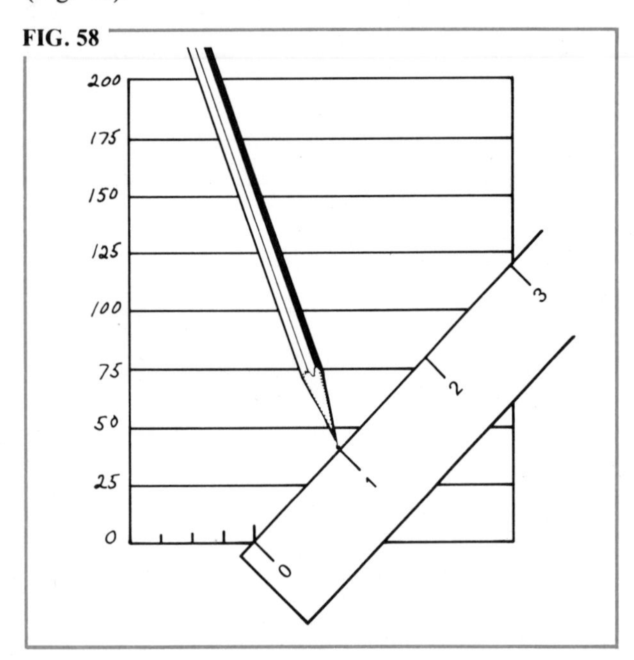

Set up your T-square and triangle, aligning the triangle with the tick mark. Draw a light line through the base line (Fig. 59).

This establishes the width of a column. Now to unscramble all these tick marks. What you now have is a base line with all the spaces represented on one side and all the columns on the other. They have to be rearranged. Here is how to do it. Remember, this is not a gimmick; it is a legitimate, accurate trick of the trade.

Take a small piece of paper (about 4″ long or so) with a straight edge on one side, position it on the base line, and tick off the column width and one space mark to the left of the column (Fig. 60).

What you have created is a ruler of your own that represents the space between columns as well as the width of a column (Fig. 61).

Erase the marks on the base and repencil the line. Take your paper ruler and set it up on the base line, with the first mark at the beginning of the base. As a general rule the lower left corner of a chart (the beginning of the base line) is usually labeled 0. Tick off the two remaining lines (Fig. 62).

You have just marked the width of a space and the first column. Move the paper ruler along the line until the first mark lines up with the end of the first column (Fig. 63).

This establishes the second column. Move the paper again, repeat the marking, and you have created the third column. The remaining space to the end of the line should be equal to the other spaces (Fig. 64).

This method is only as accurate as you are in

FIG. 59

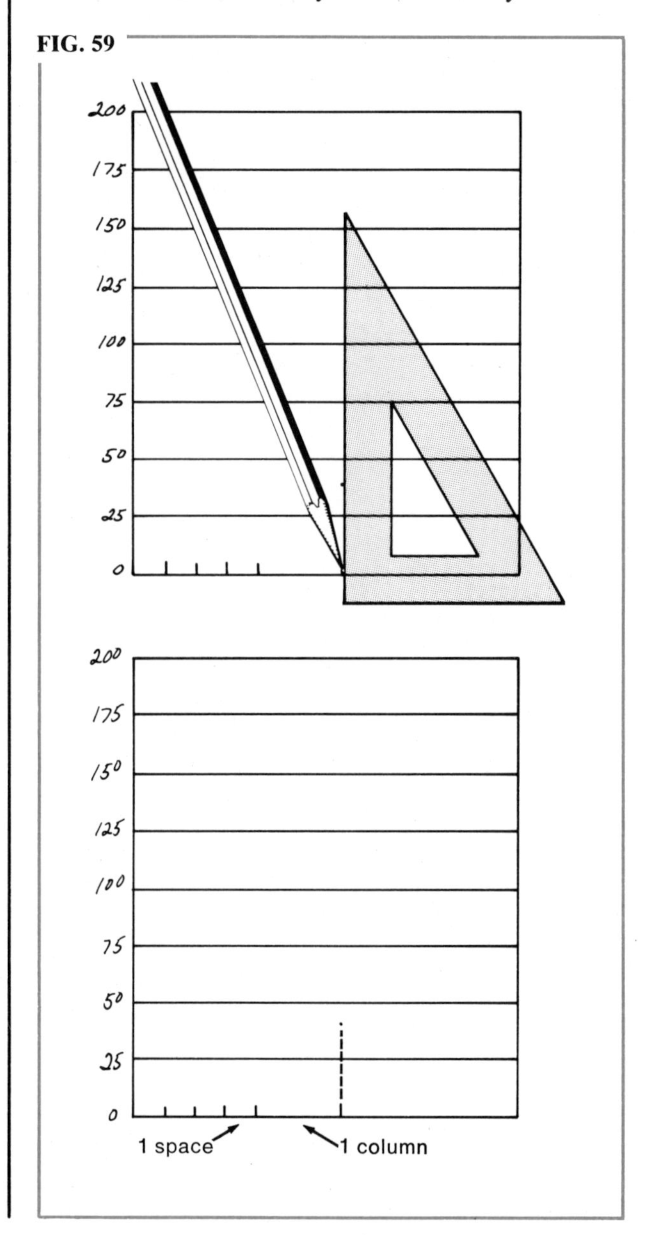

FIG. 60

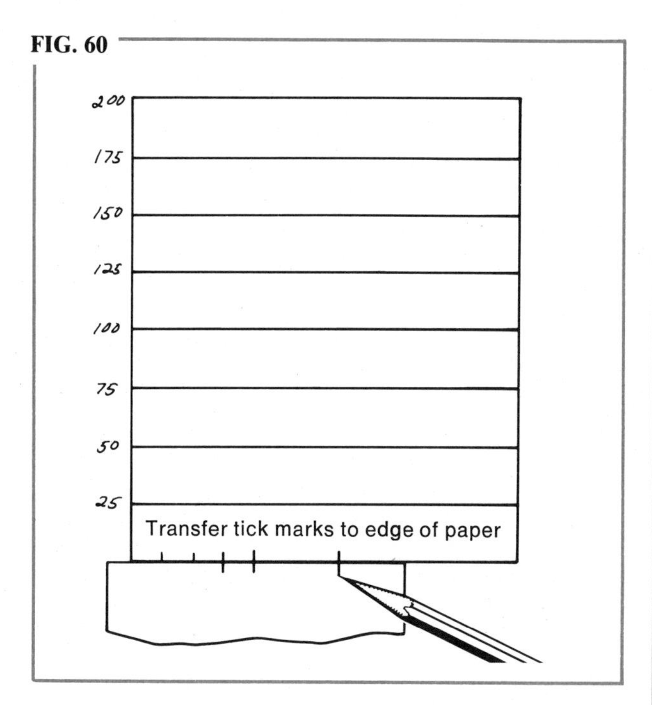

FIG. 61

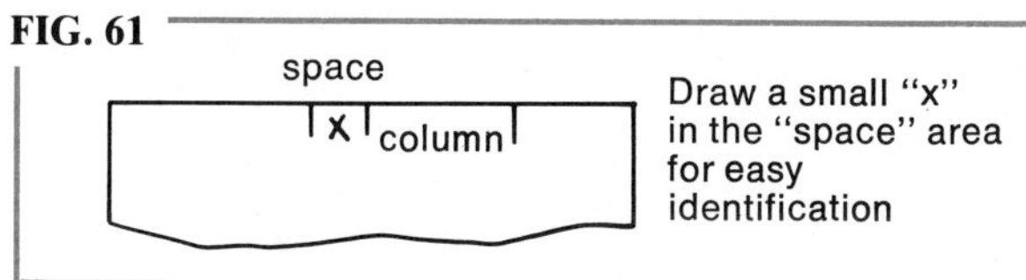

FIG. 62

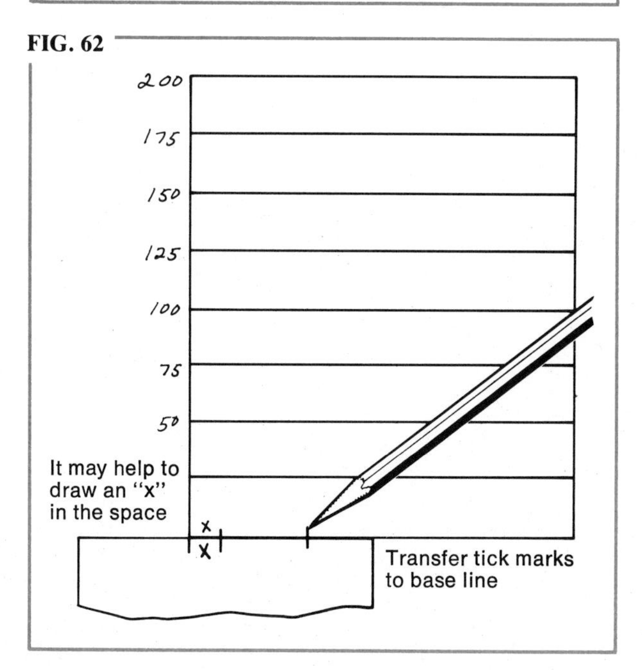

FIG. 63

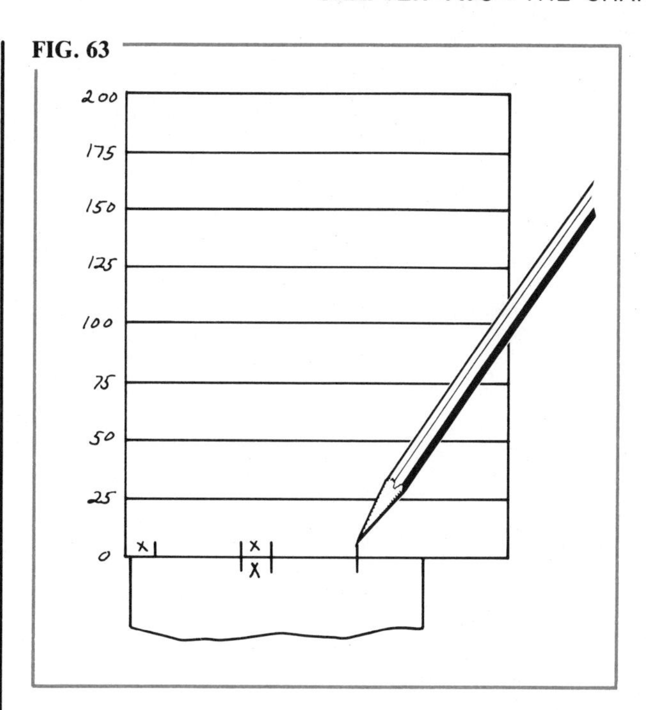

FIG. 64

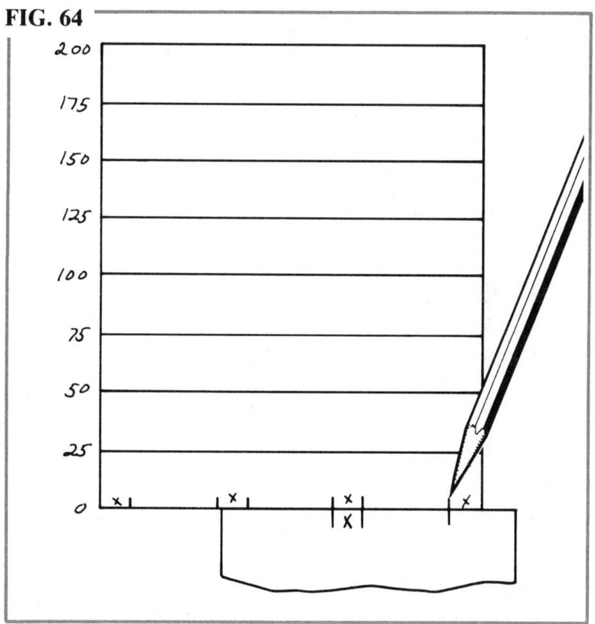

its execution. If you find that the last tick mark is slightly off (often just the distance of the thickness of your tick mark), it means you weren't as precise in measuring as you should have been. Just do it over until you get it right. Remember: sharpen that pencil to a needle point.

Set up your T-square and triangle and draw perpendicular lines through each tick mark. Label each column to avoid confusion later (Fig. 65).

FIG. 65

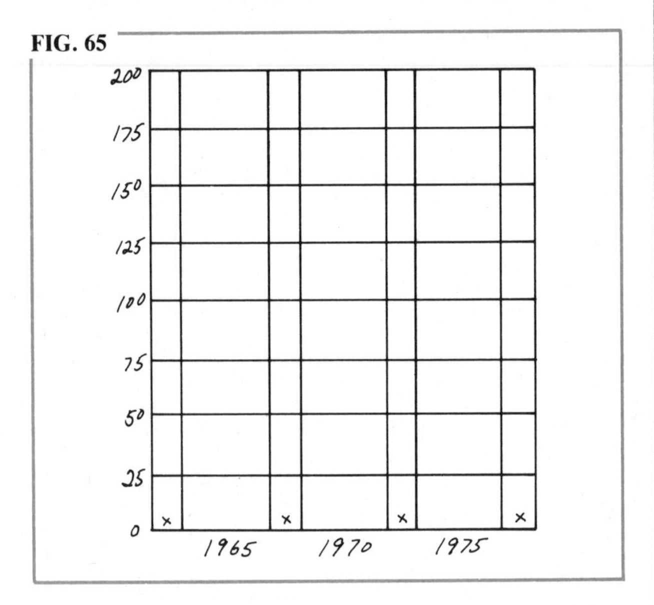

Here you have the basic structure or skeleton of the chart. The next and final step is to indicate the proper height of each column.

At this point in its preparation it should be apparent that all those diagonal guidelines and tick marks, criss-crossing the grid lines and the column lines, could look quite confusing. In more elaborate charts they could really drive you to distraction. It would be wise to develop certain habits as you go along, such as labeling everything as you draw it, as well as erasing certain areas of guide lines from inside the columns. This will help you identify the columns more readily, giving you a better overall picture of the chart.

Establishing the correct height of each column is the final, crucial step. This could be considered the most difficult of all because you must work with a small space, and great patience is required for its preparation.

However, before going any further, let us get one thing out of the way. That is the "eyeball" method. There will be times when preparing a chart by eye is all that will be necessary. So here is a simple approach you may use.

The height of the first column is to be 140. Focus your attention on the space between the 125 line and the 150 line. The distance between each division on the grid is 25. Visually position your pencil point halfway between the 125 and the 150 lines. This represents 137.5. Move your pencil up to halfway between the center point and the 150 line, which is about 143.7. Slightly below that point is just about 140. That point is the height of the first column. Position your T-square on that point, draw a line across to close off the top of the first column, and label it 140 (Fig. 66).

FIG. 66

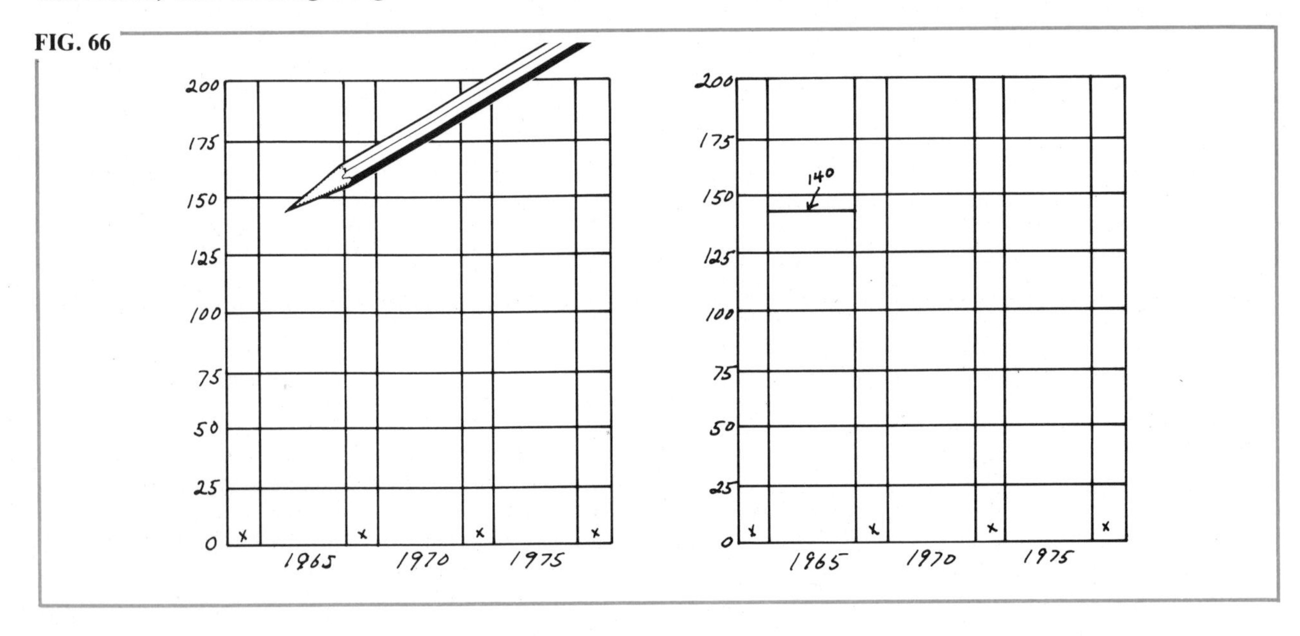

Repeat the same approach with the next column. The 1970 column is to be 172. Work between the 150 line and the 175 line. Place a light mark in the center (at about 162); then a mark between the center and the 175 line (about 168). Halfway between that mark and the 175 line is approximately 172. Close enough.

Approach the third column in the same manner (Fig. 67).

FIG. 67

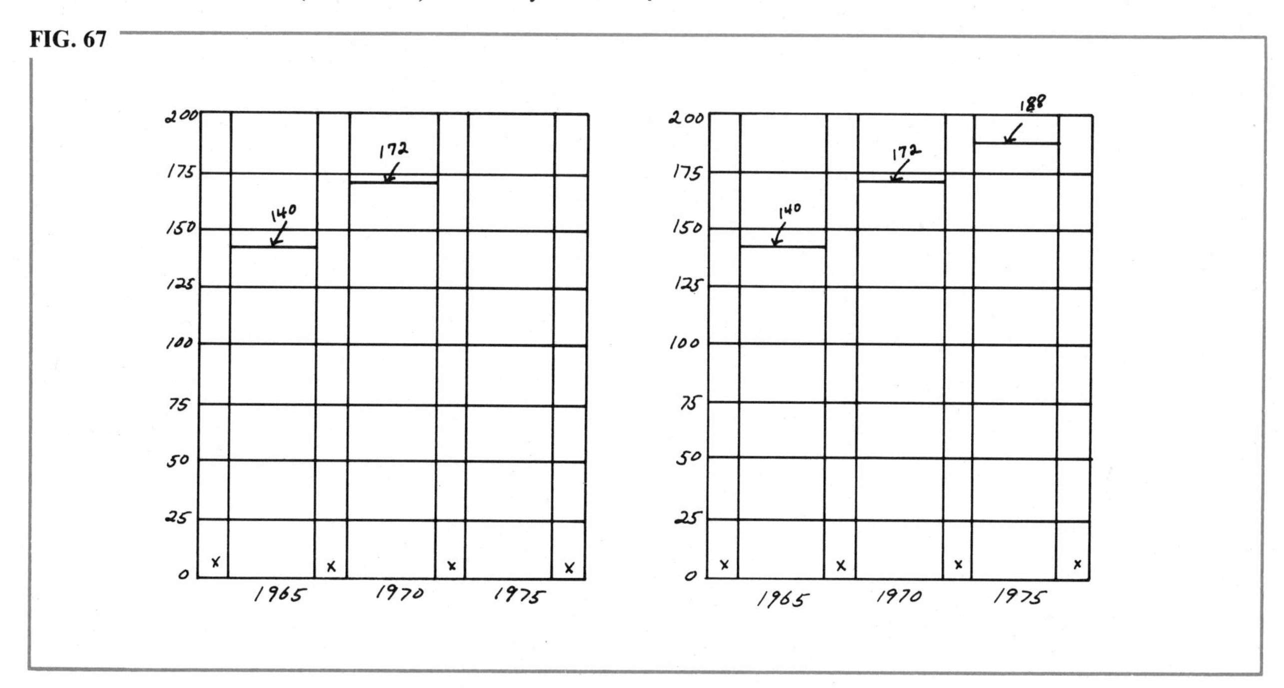

This method of plotting is fine if absolute accuracy is not essential. However, it is the last time it will be mentioned in this text. From this point on we will be concerned with accuracy.

Once again, back to serious business. Another term to be used here is *plot*. The act of calculating a point on a chart is called *plotting* that point. In order to plot a point, you will need to prepare a *grid* scale. Usually a scale represents the entire height of the chart, whereas the grid scale represents one segment of the scale broken down into smaller increments.

Your grid scale can be prepared between any two grid lines on the chart. But it is probably less confusing to do it between the 0 and the first grid line.

Take a clean piece of white English vellum about 3″ or 4″ square, with at least one straight edge. Place the straight edge of the vellum along the scale line, visually centering the first segment (0–25) as shown in Fig. 68, and tape it at the top and bottom so that it cannot shift (Fig. 68).

FIG. 68

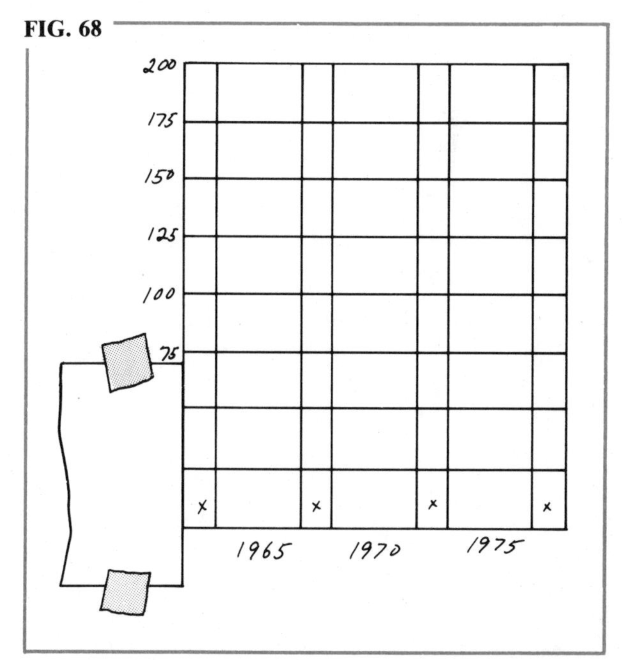

With your T-square and a 6H pencil, very carefully extend the base line (0) and the first grid line (25) onto the vellum (about an inch or more), as in Fig. 69.

FIG. 69

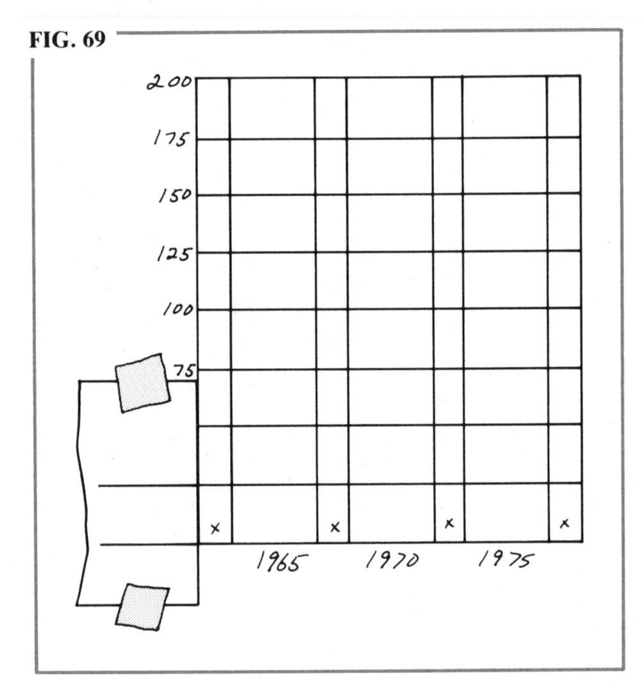

For the next step you might try using your engine divided ruler. This three-sided ruler displays 12 inches along each of its six edges. Each edge shows the inch equally divided into a given number of segments and is accordingly labeled 10, 20, 30, 40, 50, 60. You can actually use any ruler, but you will generally find that the requirements of most grid scales determine which one to use. The narrower the segment, the more difficult this becomes.

Ordinarily a segment represents uniform increments, generally in quantities of 5 or 10, such as 5, 10, 15, 20 or 10, 20, 30, 40, 50. If applicable, the increments may be 1, 2, 3, 4, 5 or 2, 4, 6, 8, 10. This is usually indicated by the width of the segment you will be working with.

Let's start by dividing the segment on the vellum into 5 equal increments. Use the 10 divided side of your engine divided ruler and place the 0 mark anywhere far to the left on the base line. Pivot the ruler until the ½″ mark touches the top line. Pencil a light line along the edge of the ruler and tick off the 5 marks between 0 and ½″ (Fig. 70).

FIG. 70

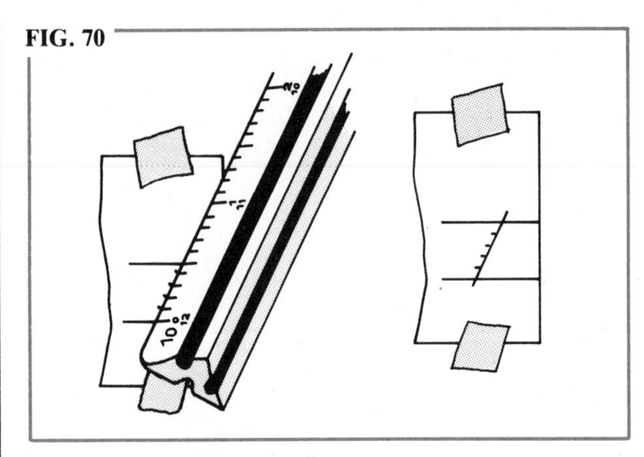

Using your T-square, draw horizontal lines through each point along the diagonal. Draw each line with as much precision as possible, being certain that each is drawn off the straight edge of the vellum (Fig. 71).

FIG. 71

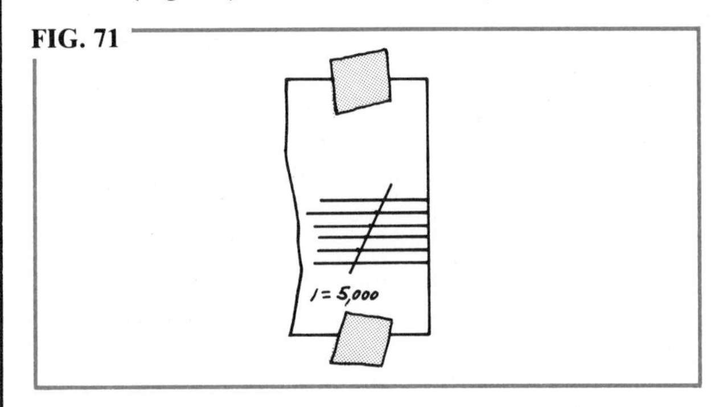

Label the vellum 1 = 5,000. This is another safety measure to remind you what the scale represents while you are plotting figures. Remove the vellum. Observe that you have created a scale for one grid section representing a total of 25,000 people. Each grid line equals 5,000 people.

You need to find 140 on the first column. Place the straight edge of your grid scale along the left vertical of the first column, positioned between the 125 and the 150 grid line (Fig. 72).

Count off each grid line from the 125 mark until you have reached 140. Place a mark at the 140 line. Remove the grid and construct a line across the column with your T-square (Fig. 73).

Label that line lightly in pencil "140." Labeling as you progress is a habit you should start to develop immediately. Once you've worked with complicated charts, you'll understand why. Thus you have accurately, and without question, plotted

FIG. 72

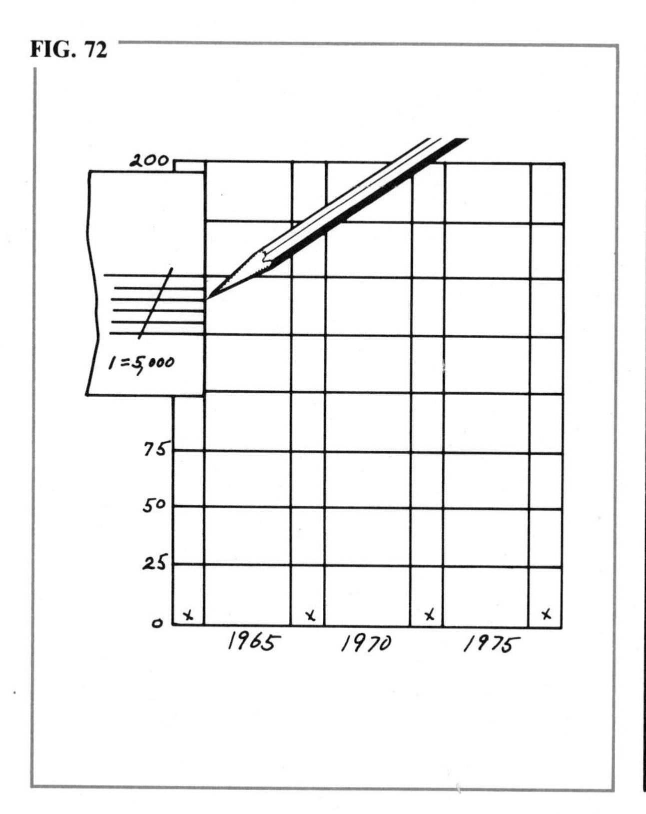

FIG. 73

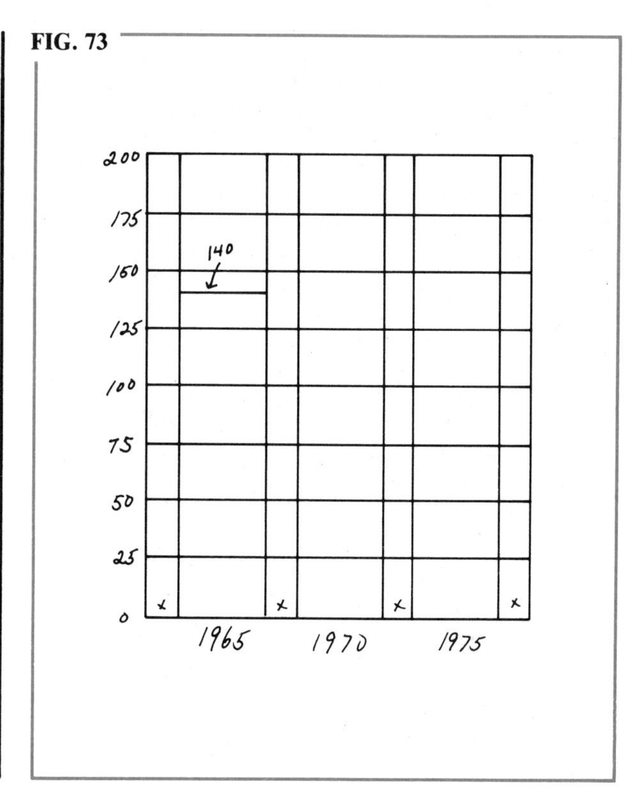

FIG. 74

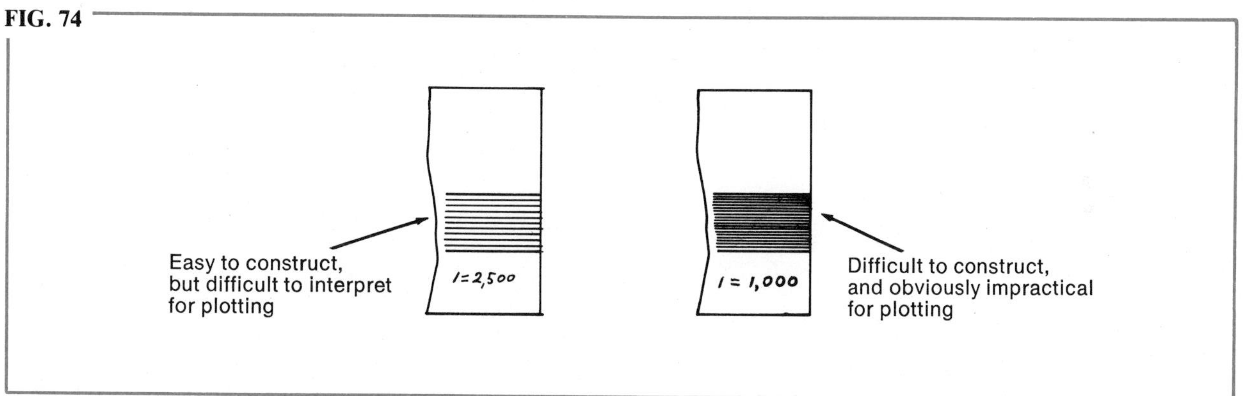

the height of the first column.

Let's study the grid for a moment. It may seem that we have arbitrarily chosen to divide the grid into 5 increments. We might have tried 10 equal parts. Each increment would then be 2,500, making it somewhat confusing to calculate the plot figures. You could, of course, attempt to break the grid down into 25 equal increments, but the lines would be so close together that you might find it impossible to construct (Fig. 74).

All this is relevant to the requirements of plotting. The more intricate the number you must plot, the more increments you'll need on the grid to help you locate it. For instance, the number 140.76 would be more difficult to plot than 140. The difficulty would be even greater if the next number to plot were 141.90. The chart would have to show the difference, no matter how slight. Thus there is no set rule as to how intricate you must make your grid.

In order to plot the next column, follow the same procedure. Align the grid on the edge of the second column between the 150 and 175 grid lines. Count off each grid line, 155, 160, 165, 170, . . .; the 172 mark is less than halfway from 170 to 175. This can be found by eye. It is so close it could hardly be disputed. Draw a line across at that point to close off the column and label the line "172" (Fig. 75A).

For the third column, work between the 175 and 200 lines. Count off each grid line, 180, 185, 190.

Draw a line across at the 188 point and label it; you have completed the chart (Fig. 75B).

It's like everything else—it's easy when you know how.

FIG. 75

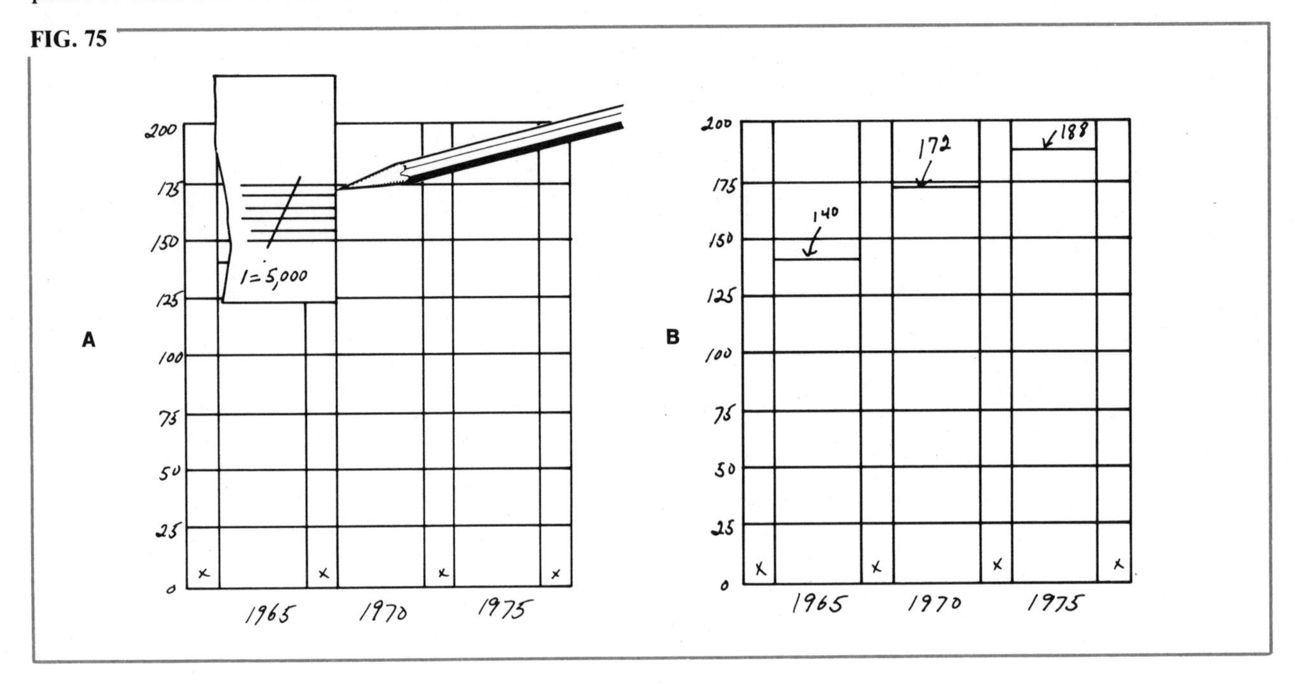

Chapter Three
Application

Having been introduced to the principles of diagonal projection and imaginative use of a ruler, you are equipped to prepare almost any kind of chart.

Try what you have learned on a bar chart. The specifications are as follows:

A sales growth comparison in millions of dollars over a period of 11 years.

The dates and figures are:

1968	$1,324	1974	$1,916
1969	$1,896	1975	$1,705
1970	$2,237	1976	$1,409
1971	$2,215	1977	$1,610
1972	$2,190	1978	$1,820
1973	$1,964		

The overall dimensions are $5\frac{1}{4}''$ wide by $3\frac{5}{8}''$ high.

The thickness of each bar is to be $\frac{1}{4}''$.

The space between bars is to be determined.

The highest number is $2,237,000. The full width of the chart will represent $2,250,000 with a scale of 25 increments of 100,000 each.

Start by preparing a rough sketch of the chart according to the specifications (Fig. 76).

One aspect of the specifications that is frequently overlooked but should be considered is the area to be occupied by the surrounding type. For example: Does the $5\frac{1}{4}''$ include the type, or do the given dimensions represent only the actual chart?

In this chart we will not be concerned with the type area but will include the extended tick marks along the bottom. Let's start by constructing the overall dimensions. If we establish the tick marks at the bottom to be $\frac{1}{16}''$ long, we must draw a guideline up $\frac{1}{16}''$ from the bottom edge of the given dimensions.

The specifications require 11 bars (one for each year), each $\frac{1}{4}''$ deep. Counting down from the top (or up from the bottom), tick off eleven $\frac{1}{4}''$ marks (Fig. 77).

To refresh your memory, the $2\frac{3}{4}''$ mark represents the total of the 11 bars; the remaining length represents the total of all 10 spaces between the bars. What you need to determine is the width of one space. Therefore, draw a horizontal line across the chart at the $2\frac{3}{4}''$ mark. Place your ruler in the lower left corner and pivot it until 10 equal segments of your ruler fall between the two horizontal lines ($10 \times \frac{1}{8}'' = 1\frac{1}{4}''$). Tick off one $\frac{1}{8}''$ mark (Fig. 78A). Align your T-square with that mark and extend it to the left vertical (Fig. 78B).

FIG. 76

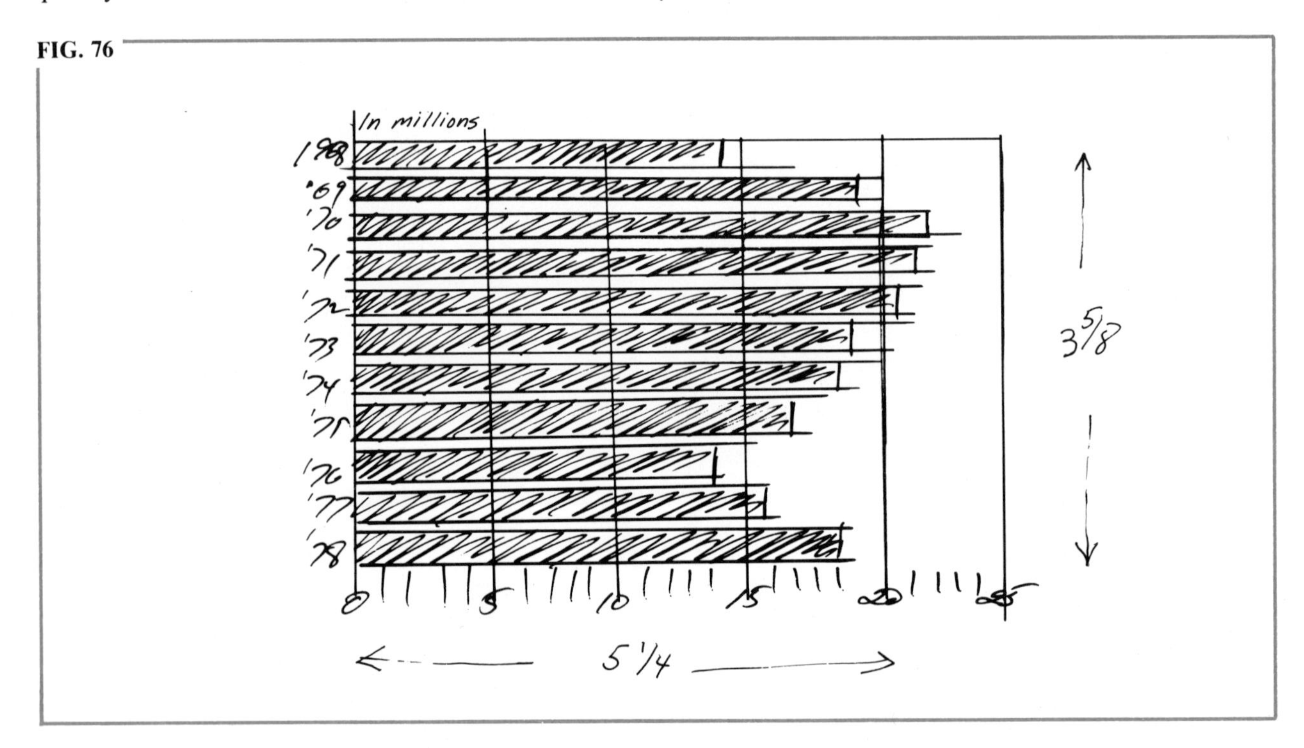

FIG. 77

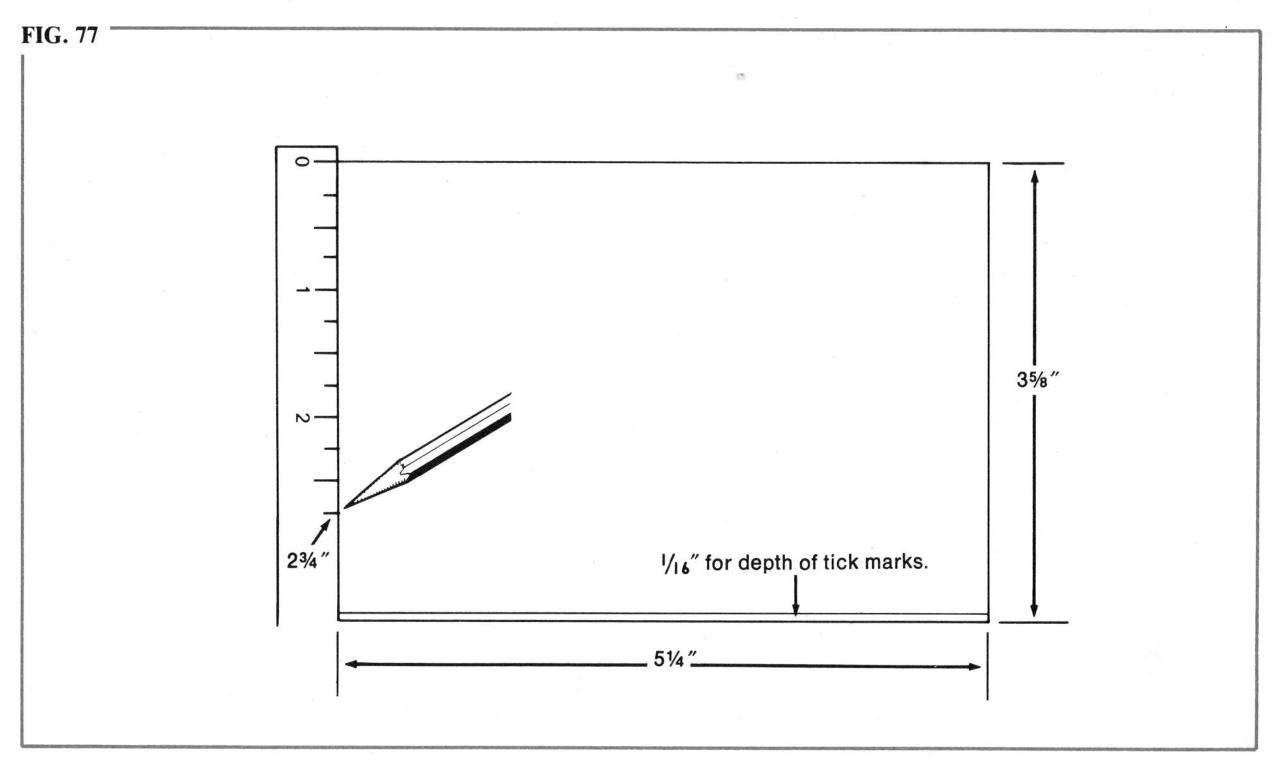
0
1
2
2¾″
1/16″ for depth of tick marks.
3⅝″
5¼″

FIG. 78

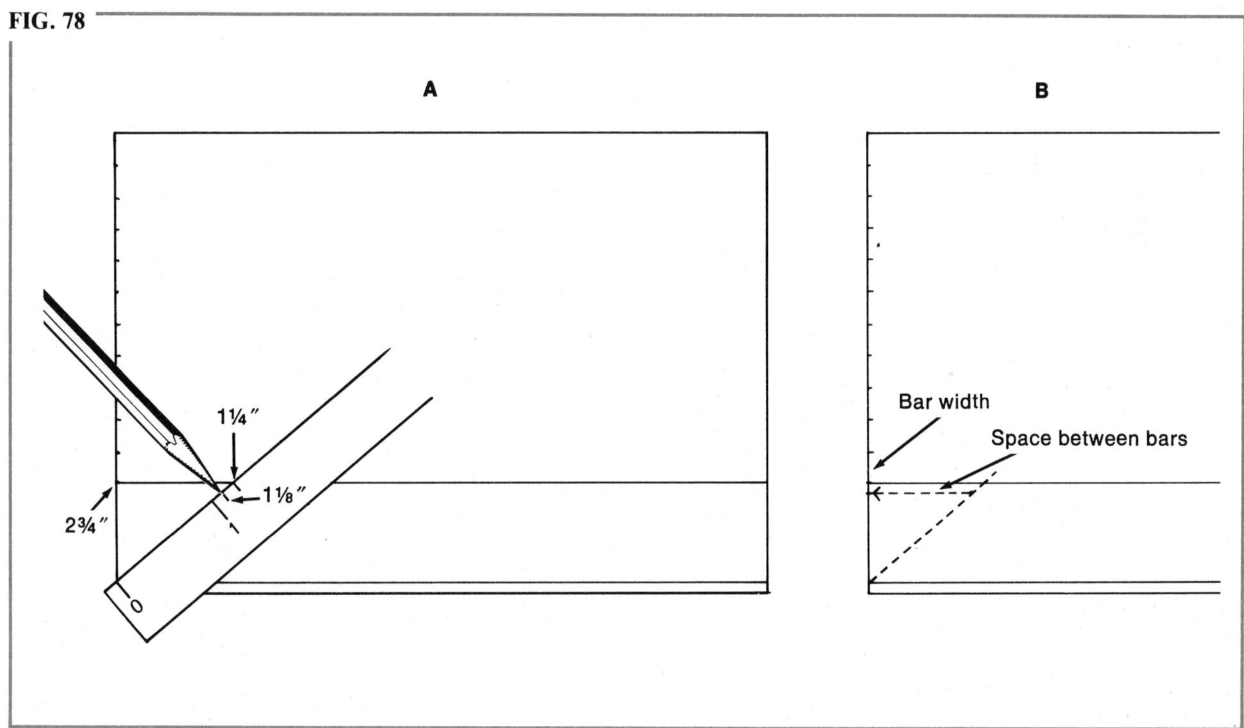
A
B
1¼″
1⅛″
2¾″
0
Bar width
Space between bars

You have established the space between bars. Place a piece of paper along the edge of the left vertical and tick off the space marks and the width of one bar (Fig. 79).

FIG. 79

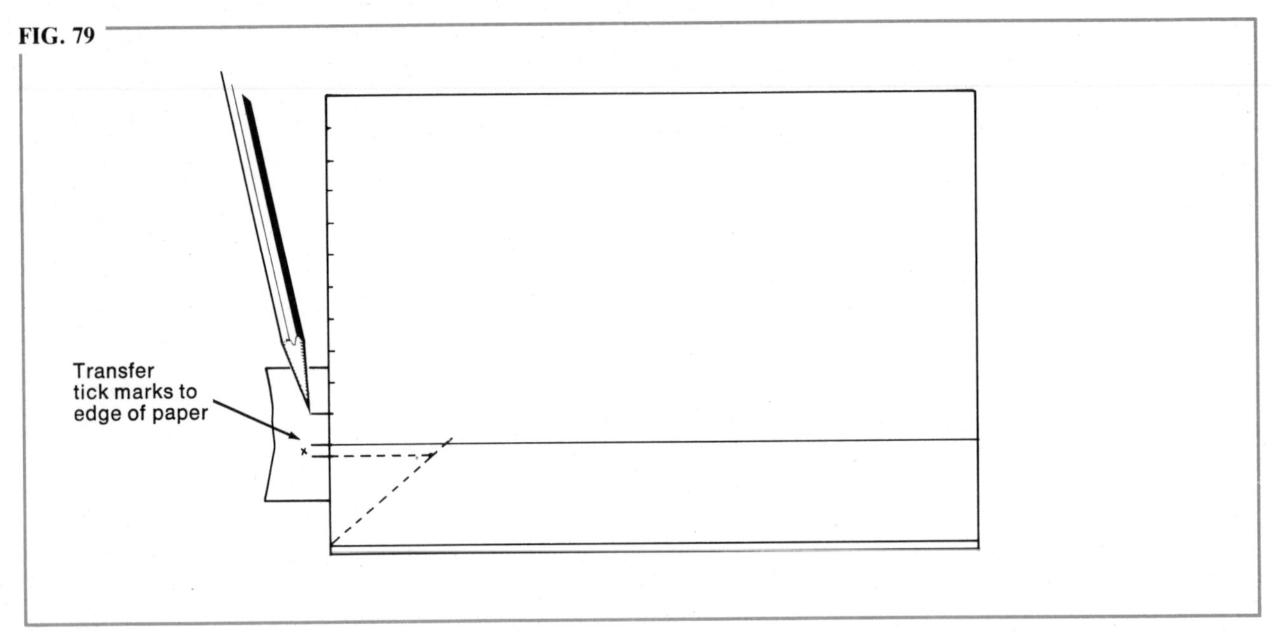

Now you are ready to construct the bars and spaces in their proper relationship. Erase all the tick marks, repencil the vertical line, and line up the paper scale with the top of the chart. Tick off the bar and space lines; move the paper down and line up the first bar mark with the space mark; tick off the bar and space marks again; repeat this procedure, moving down the vertical until you have marked all 11 bars and 10 spaces (Fig. 80).

FIG. 80

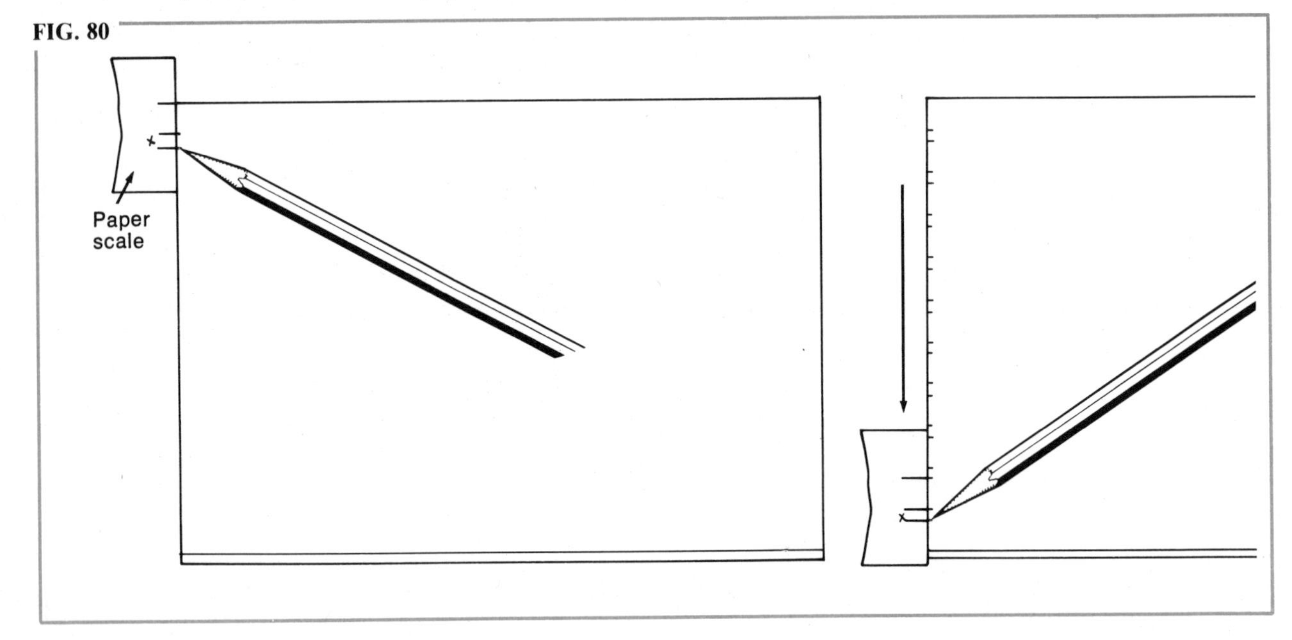

With your T-square, draw all the lines for the bars completely across. Write dates alongside each bar (Fig. 81).

FIG. 81

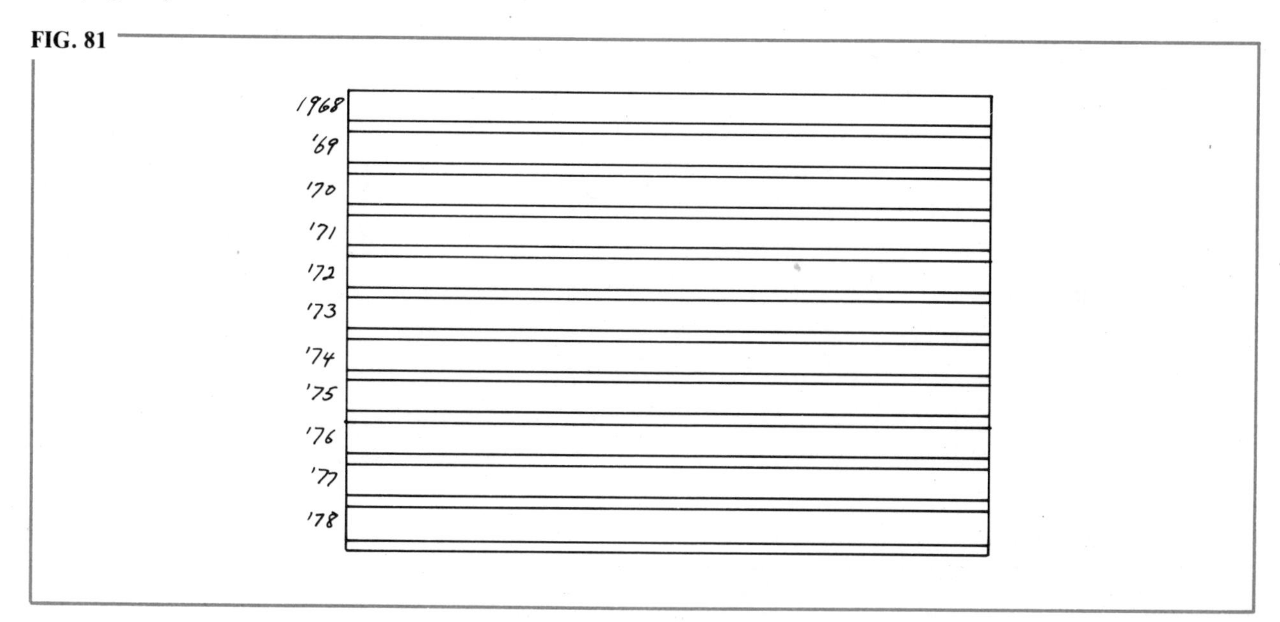

The next step is to prepare the scale and grid representing the millions of dollars. There will be 25 increments of 100,000 each. First you must determine a division on the ruler that can be easily used. The width of the chart, which is $5\frac{1}{4}$″, contains only 21 quarter inches, but there are 25 quarter inches in $6\frac{1}{4}$″. Place the 0 mark of your ruler in the lower left corner and pivot the ruler upward until the $6\frac{1}{4}$″ mark touches the right vertical (Fig. 82).

FIG. 82

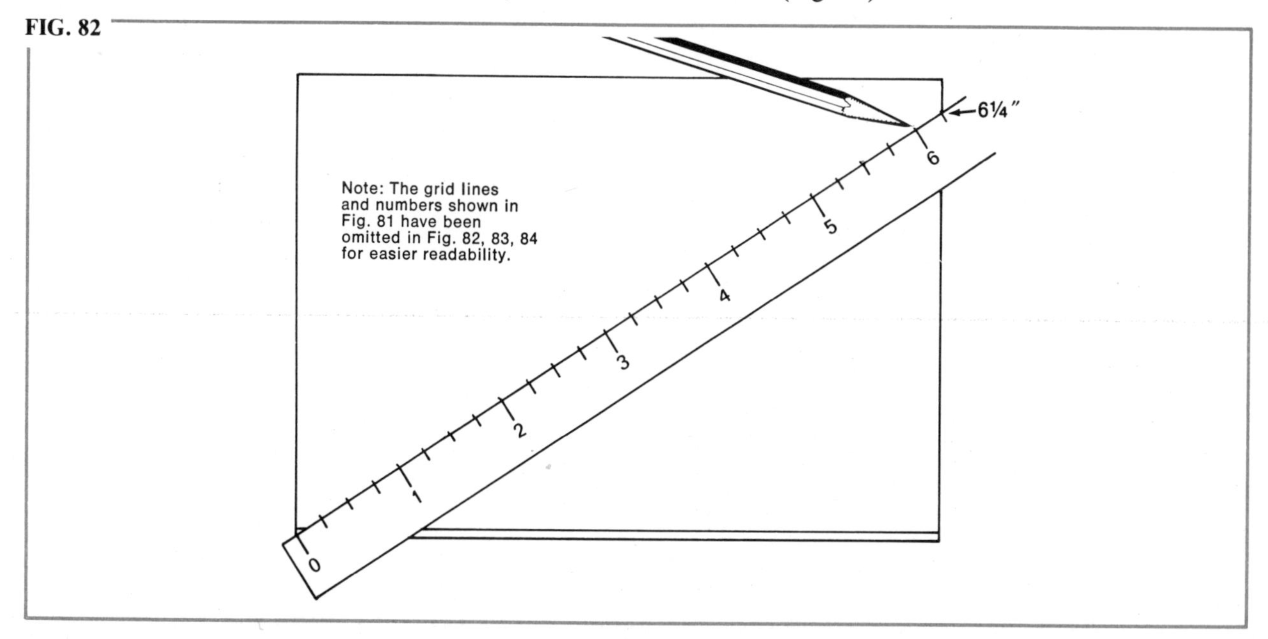

Mark off every $\frac{1}{4}''$ and draw vertical lines through each mark. Label every fifth tick mark along the bottom (Fig. 83).

For the grid scale use a piece of vellum, which is more durable than tracing paper or bond. Frosted acetate could also be used. Tape the vellum straight edge anywhere along the bottom line. With your T-square and triangle extend the lines of one segment onto the vellum (Fig. 84).

This space represents 100,000 units but is

FIG. 83

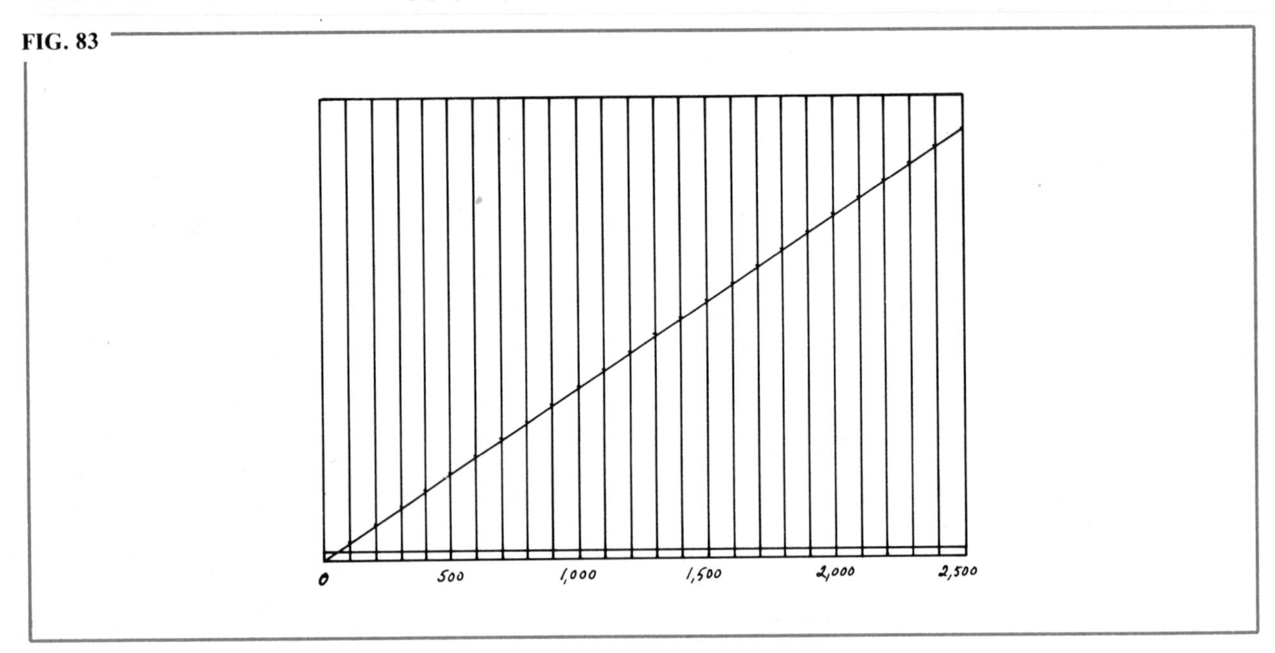

FIG. 84

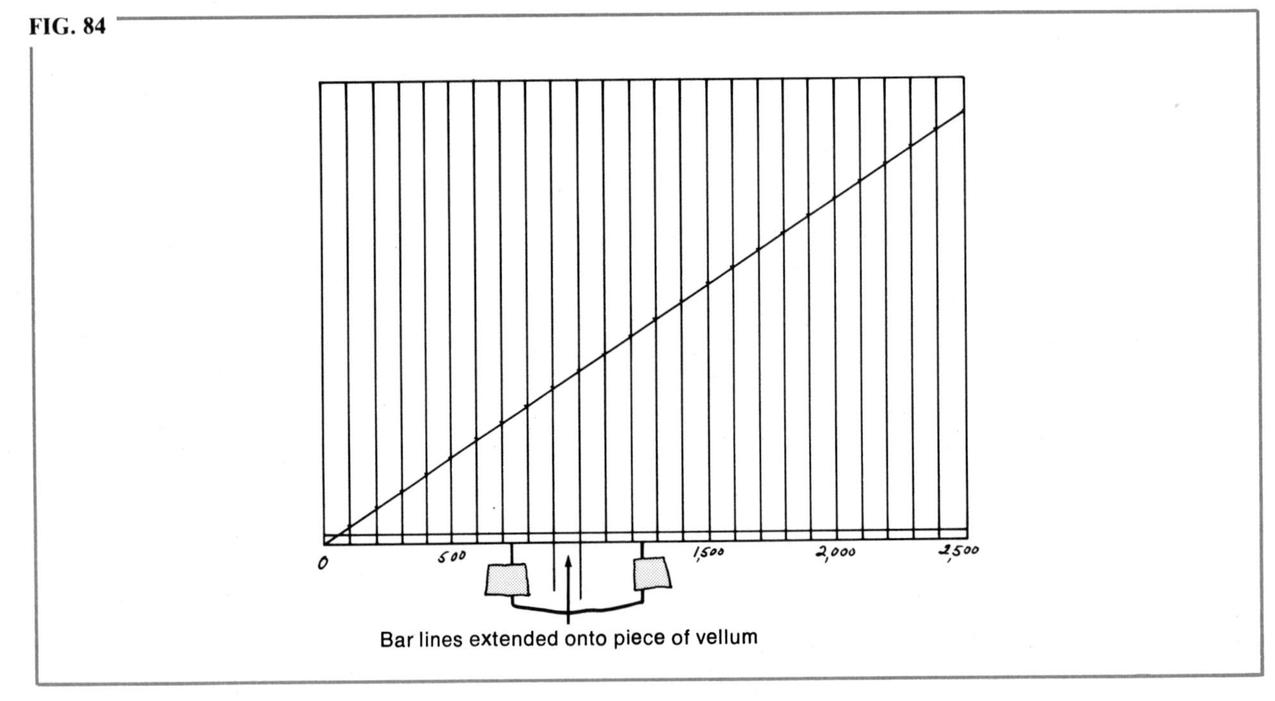

quite narrow. To divide it into 10 spaces would be an unnecessary task. The lines would be so close together that they could hardly be seen. Five spaces would be sufficient. Each space would then represent 20,000. Keep in mind that the grid could be broken down into 10 parts if the project required that much precision in the plotting. But generally this is unnecessary.

FIG. 85

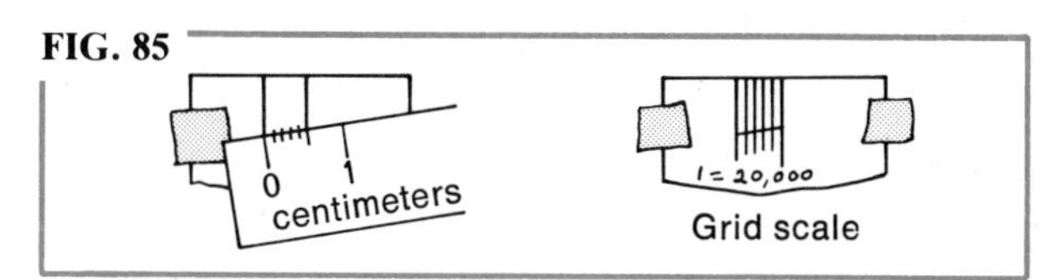

Because the space to be divided is narrow, you will need to find 5 small segments on your ruler. Rather than use the engineer's ruler, let's try the metric ruler this time. Notice that the grid segment measures just about ½ centimeter (5 millimeters). You need only angle it ever so slightly to touch both lines. Tick off each millimeter, draw vertical lines through each mark, and tick off the

FIG. 86

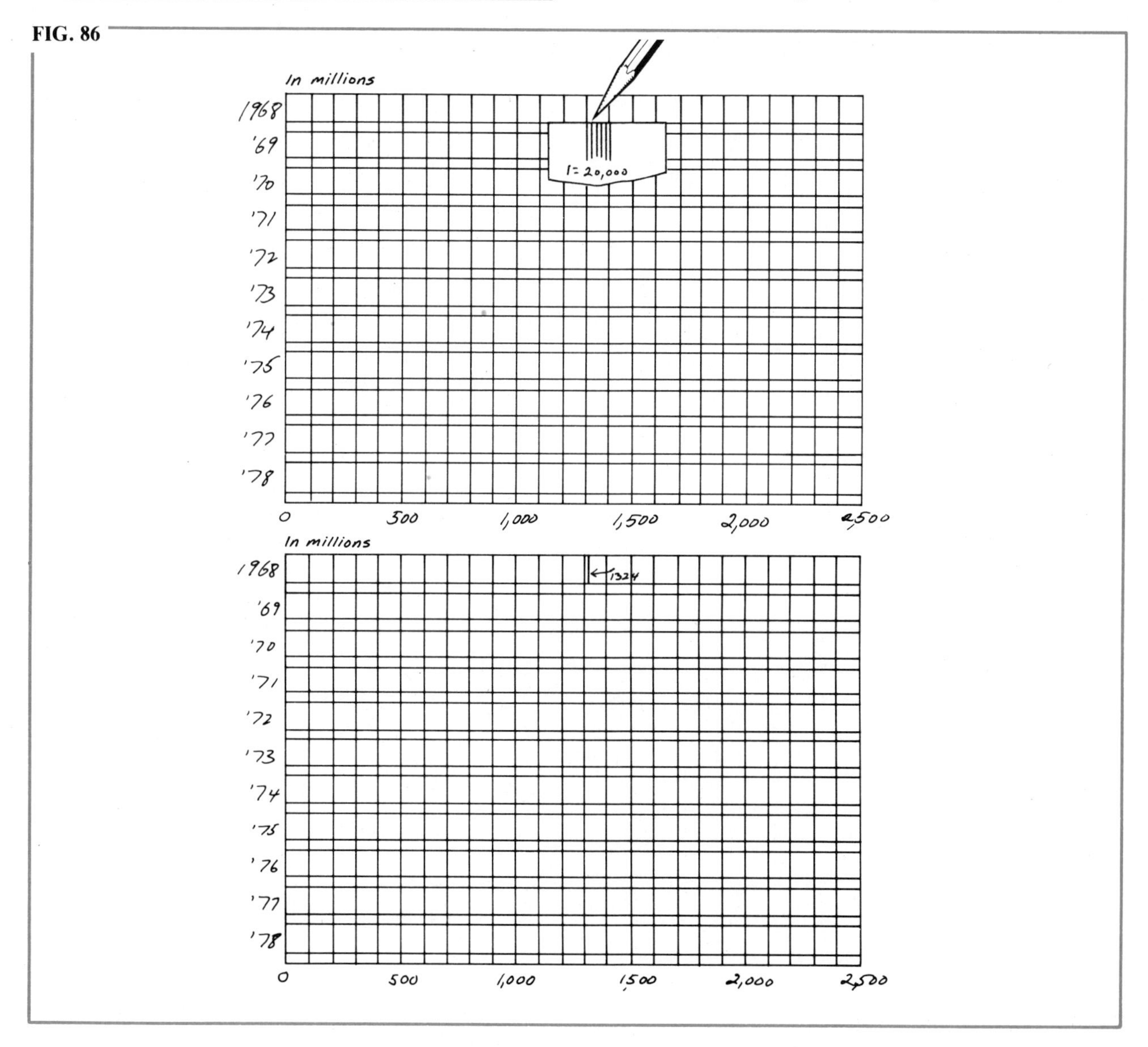

edge of the vellum; now you have your grid scale (Fig. 85). It's that simple.

Try it again with a regular ruler using five $\frac{1}{16}''$ marks. The results will be identical. Any ruler will do the job.

You are ready to plot the chart.

Start at the top bar for 1968. The figure is 1,324,000. Place your grid scale between the 1,300 and 1,400 lines. The first space mark on the grid equals 1,320. Thus 1,324 would be very slightly to the right. This proportion of the plot must be done by eye. The purpose of the scale is to bring you as

FIG. 87

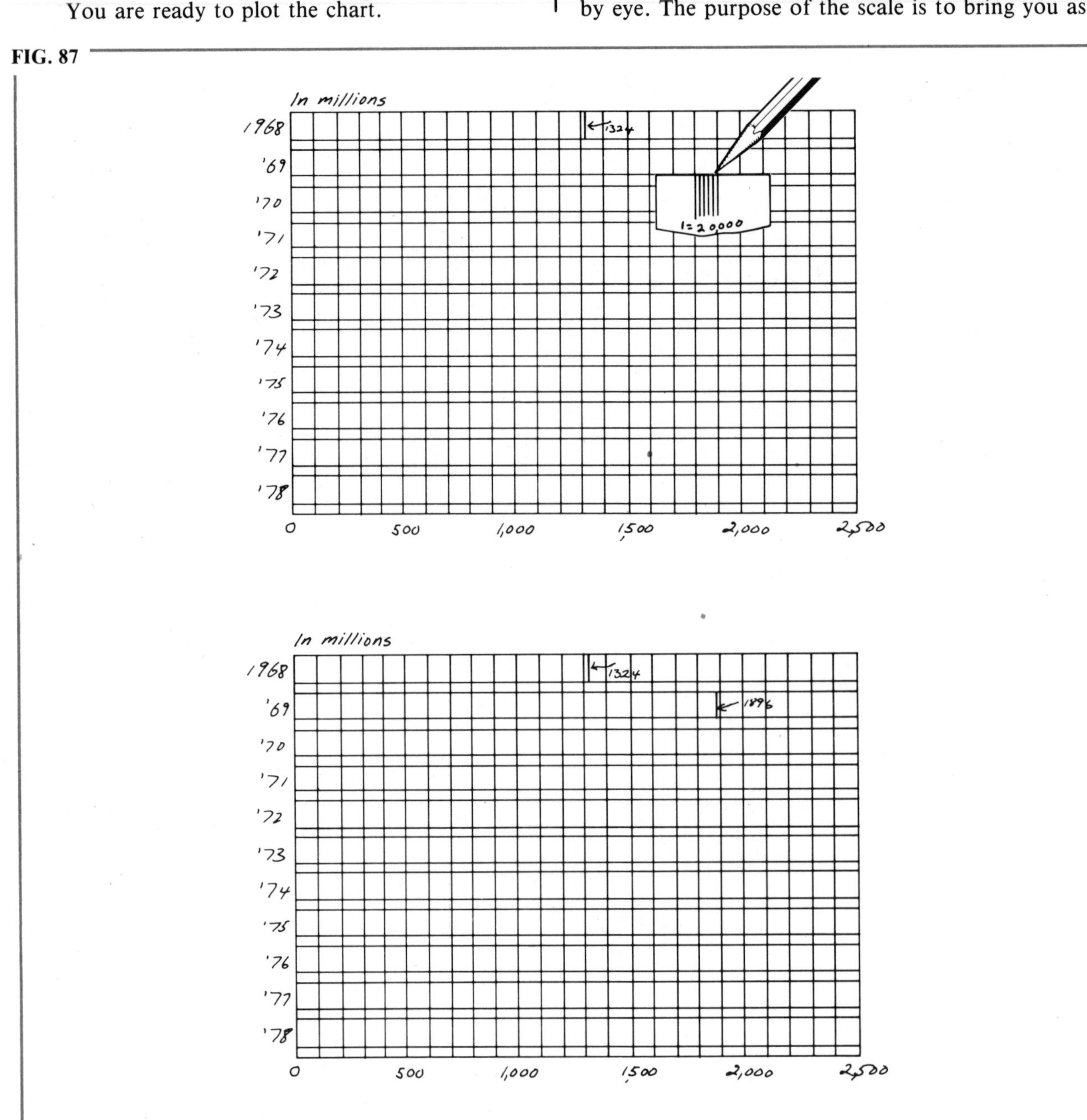

close to a plot point as possible, if not exactly to the point. Draw a vertical line at that point to end the first bar. Label it 1,324 (Fig. 86).

Place your grid scale between the 1,800 and 1,900 marks along the second bar. The fourth mark on the grid scale represents 1,880, and 1,896 would be just short of 1,900. Tick off the 1,896, draw a vertical line through it to close off the bar, and label it 1,896 (Fig. 87).

Repeat this procedure with each bar, and your chart is ready to be inked (Fig. 88).

It should be getting easier.

FIG. 88

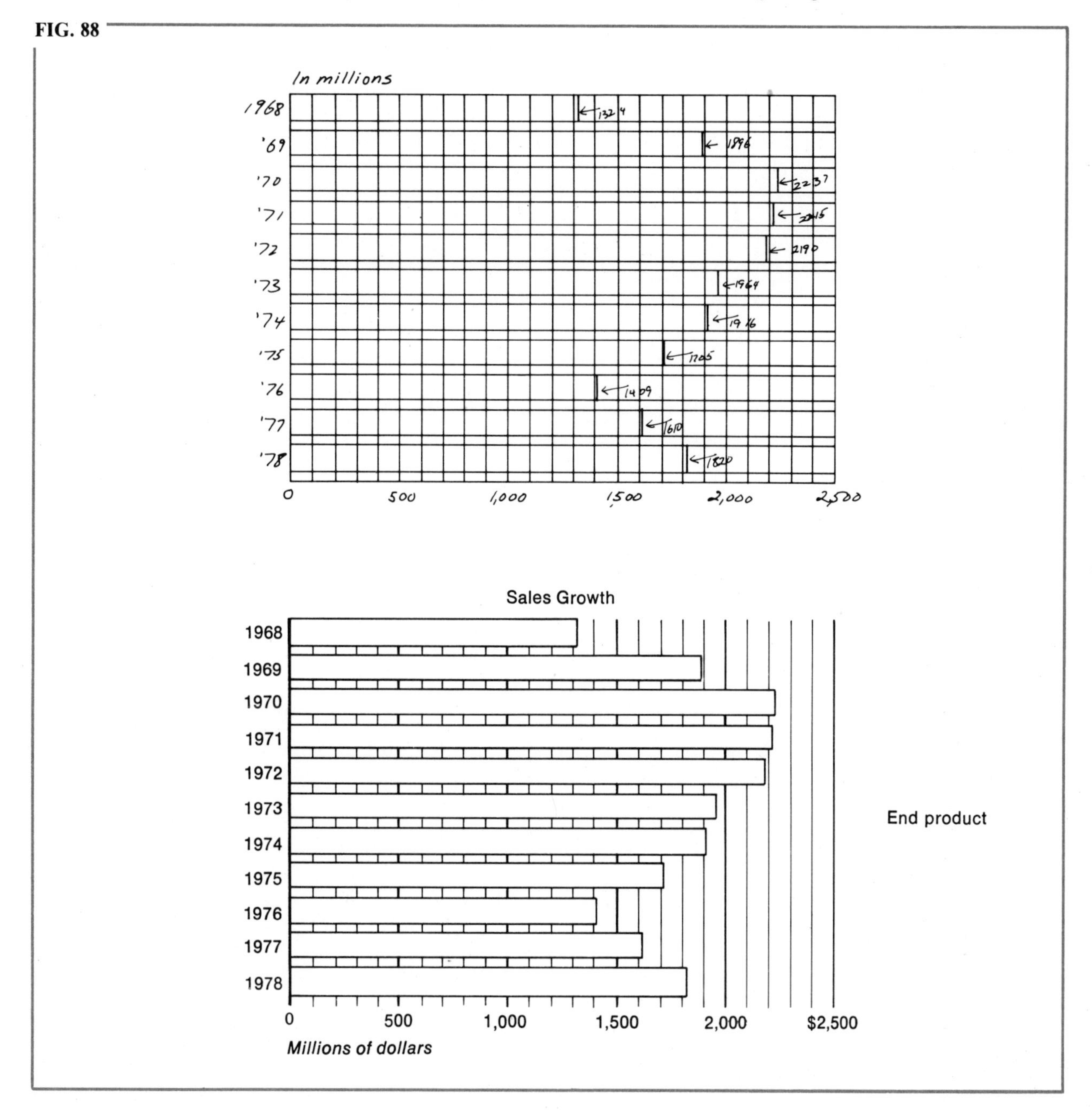

Now would be a good time to give yourself a test. Construct a chart according to the following specifications on vellum. To check it after you have completed it, place your drawing over the chart shown on page 123. If it matches, you did it right. If it doesn't, find out why!

Exercise

A horizontal bar chart showing the number of people who buy a product at various age levels. The ages and numbers of people are:

20–25 years	8,671 people
25–30 years	10,420 people
30–35 years	6,212 people
35–40 years	4,910 people
40 and over	2,500 people

Your scale will run from 0 to 10,500. It will be broken down into 11 increments. The first 10 will equal 1,000 each, and the last one will be a half space (equal to 500).

The overall dimensions (not including type) are $4\frac{3}{16}$″ wide by $2\frac{7}{8}$″ high, including a set of tick marks $\frac{1}{16}$″ long along the top as well as at the bottom.

The space between bars must be $\frac{3}{16}$″. The sequence of bars will run 20–25 years at the top bar and so on down to 40 and over.

Chapter Four
Line Charts

You will find the line chart relatively simple now that you know how to plot figures. However, because this type of chart is often used when comparing several things, you will soon discover that it can present certain problems.

Now that you are familiar with the procedure for preparing the basic grid and scale, we will eliminate the step-by-step instructions and start directly with the grid. Construct it on a clean piece of bond by measuring the chart in Fig. 89.

FIG. 89

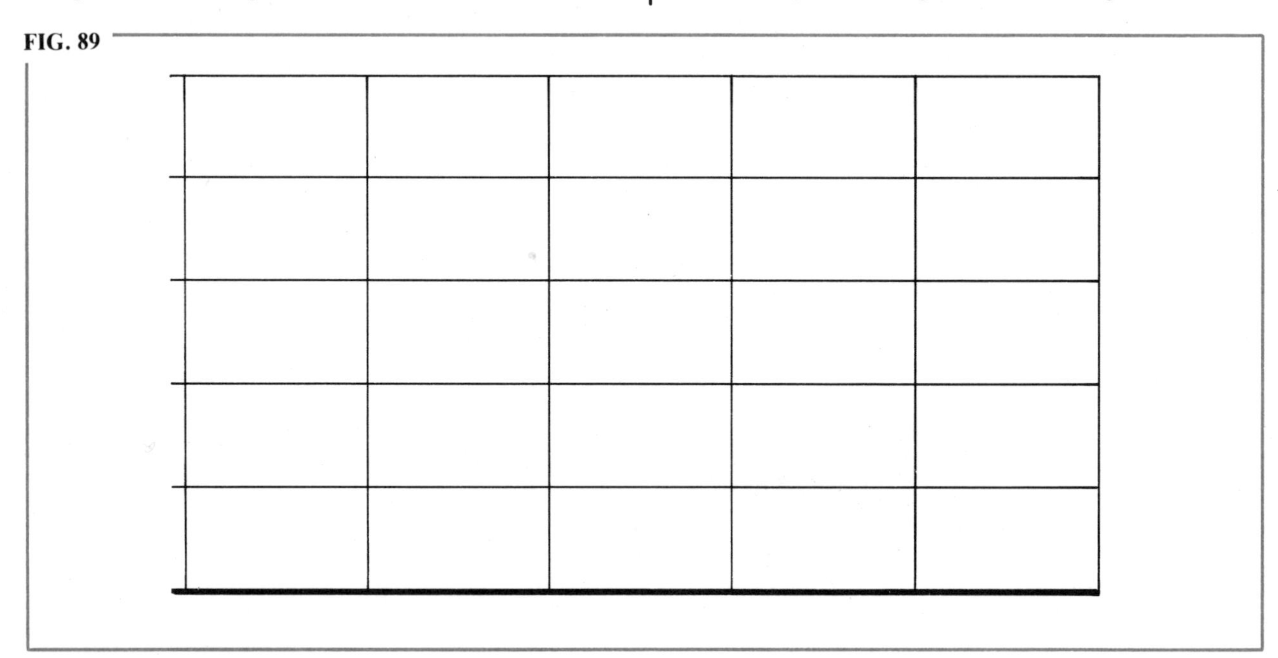

The purpose of this chart is to illustrate the growth patterns of potato production in two regions since 1955. The figures, in thousands of bushels, are as follows:

	Northeast	Midwest
1955	345,605	650,700
1960	734,915	1,140,625
1965	945,900	1,110,055
1970	1,420,005	1,210,160
1975	1,880,520	1,625,900
1980	1,995,890	2,220,400

Prepare your grid scale and plot each figure for the Northeast, placing a dot on the year line for each plot point. After all five figures have been plotted, connect them using the edge of a triangle as in Fig. 90A and 90B.

For this size chart prepare your grid scale for one 500,000 segment and subdivide it into 10 equal parts. Each part will equal 50,000. When working with figures such as these, simply count off each 50,000 and plot values between them by eye, also rounding off the number to the nearest 50,000.

For example,

345,605 = 350,000
734,915 = 750,000
945,900 = 950,000
1,880,000 = 1,900,000

With a grid this small it is impossible to plot the difference between, say, 345,605 and 350,000.

Repeat this procedure for the Midwest figures (Fig. 91A and 91B).

Here you have a simple line chart. When you prepare the final art, you will have to decide on a method of distinguishing one line from the other in addition to labeling each line. For example, one line could simply be heavier than the other. Or perhaps one line could be dashed while the other is solid (Fig. 92).

Of course, if you were preparing the chart in color (or for color reproduction), you could use different colors in addition to weights and textures of line.

FIG. 90

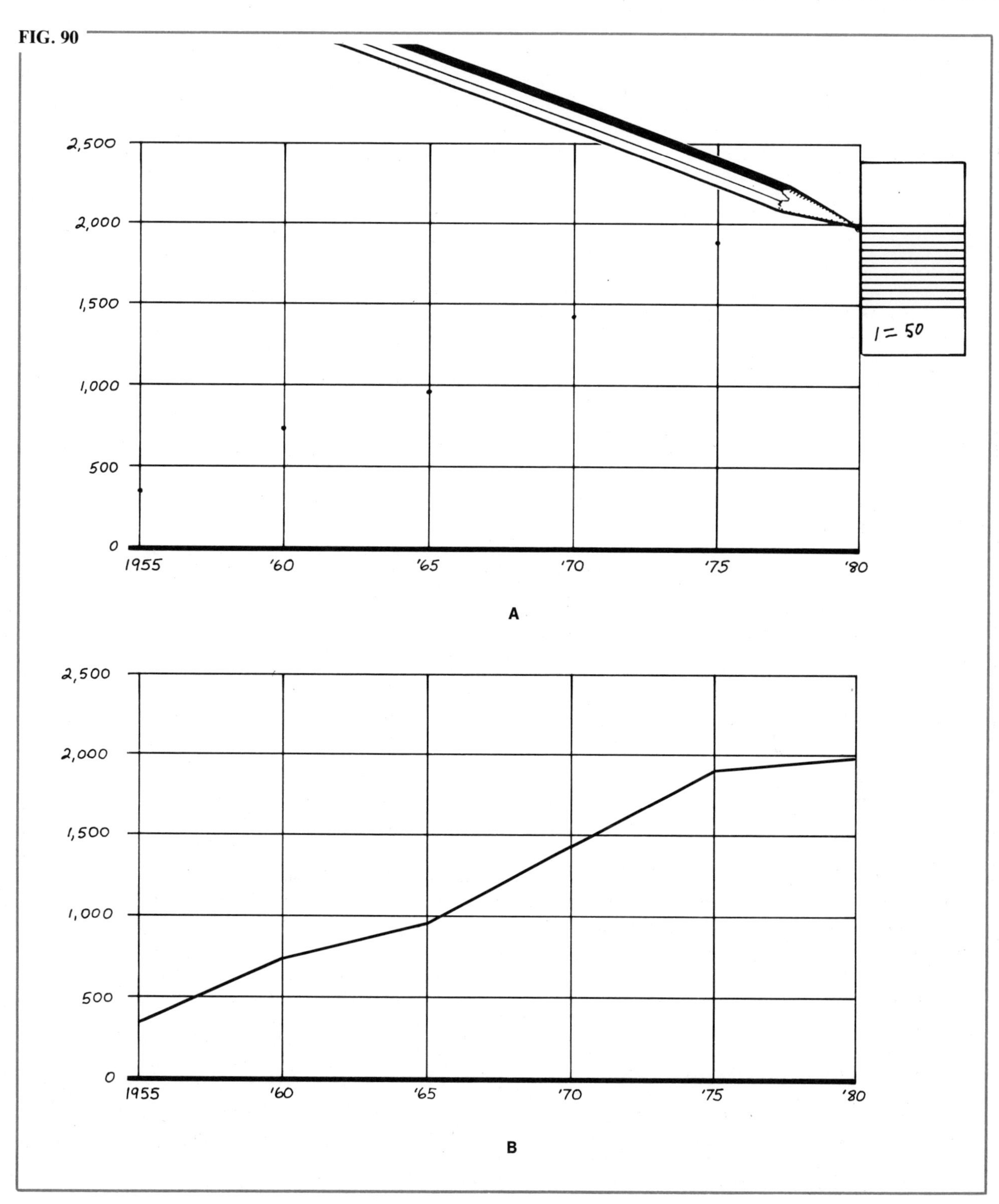
2,500
2,000
1,500
1,000
500
0
1955
'60
'65
'70
'75
'80
1 = 50
A
2,500
2,000
1,500
1,000
500
0
1955
'60
'65
'70
'75
'80
B

FIG. 91

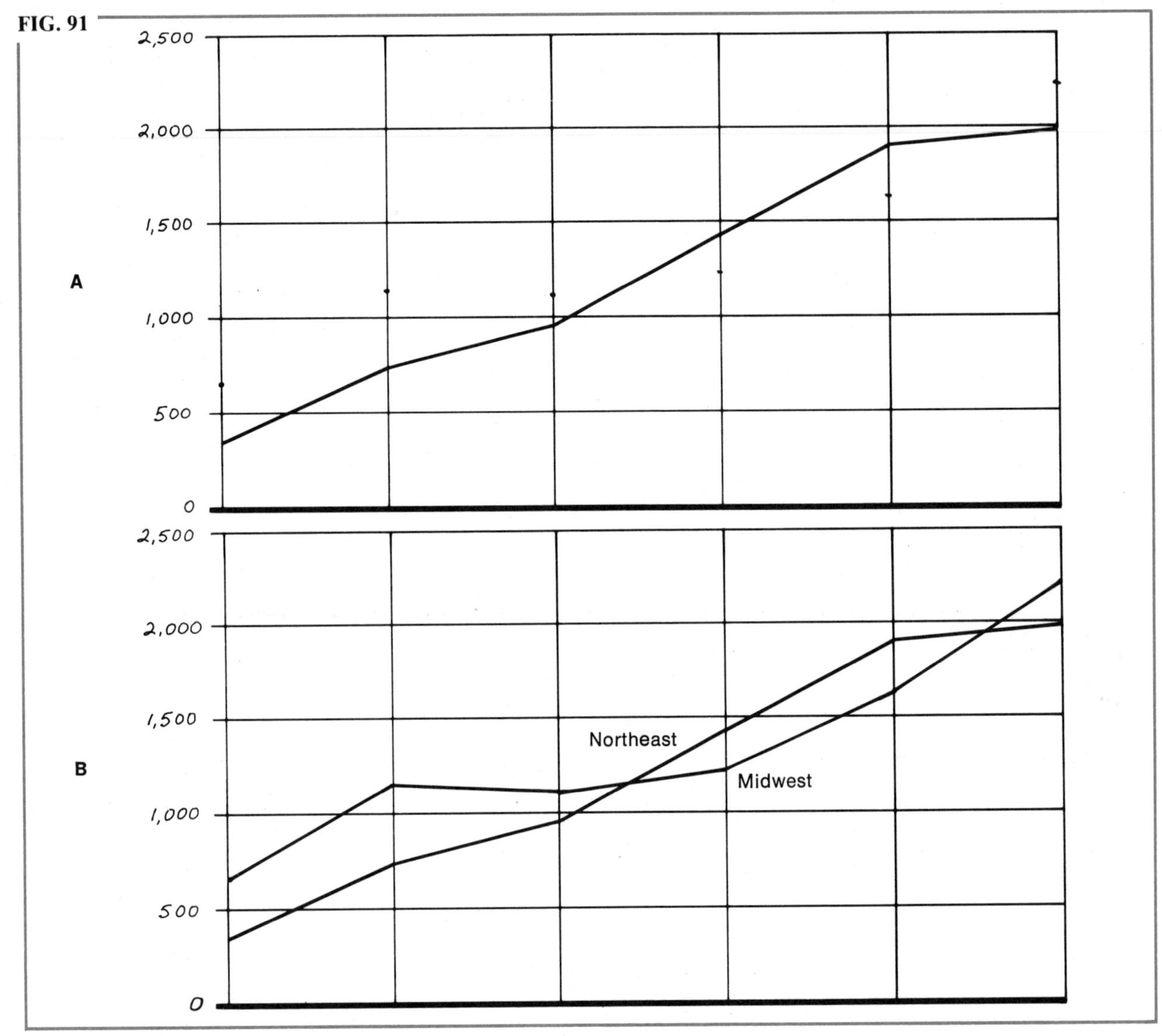

Suppose you had to show more than two comparisons. Let's add three more lines and see what problems might arise. Here are the figures, plot them very carefully:

	West Coast	Northwest	Central
1955	255,500	250,400	110,100
1960	270,450	260,590	220,305
1965	310,300	290,410	325,500
1970	490,040	305,045	505,225
1975	510,200	310,110	550,390
1980	545,360	400,450	770,775

When you have finished, your chart should look like Fig. 93.

You can see the drawback here. The lines are so close to each other that it is difficult to distinguish some of the plot points. Labeling these lines would also present a problem. You must rely on a legend (or key) to identify each line.

In a situation such as this you may be able to relieve some of the clutter by making the chart's width narrower. This might "open" the lines somewhat, as shown in Fig. 94. You would still need the legend.

FIG. 92

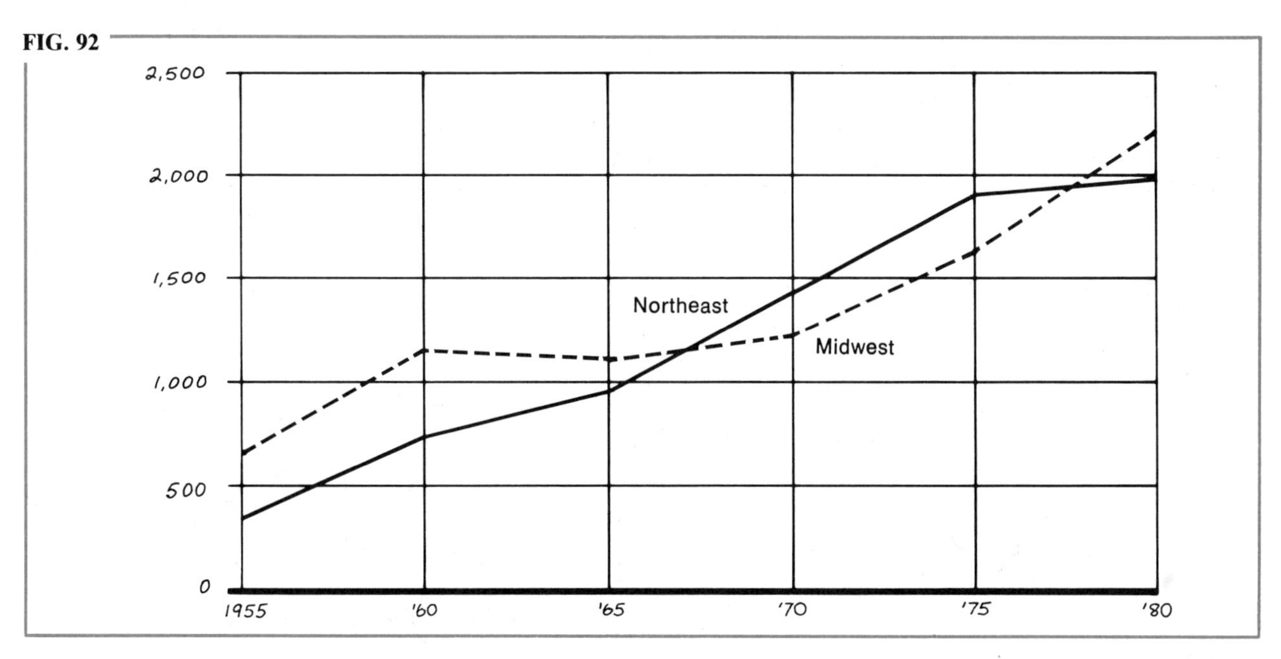
2,500
2,000
1,500
1,000
500
0
1955
'60
'65
'70
'75
'80
Northeast
Midwest

FIG. 93

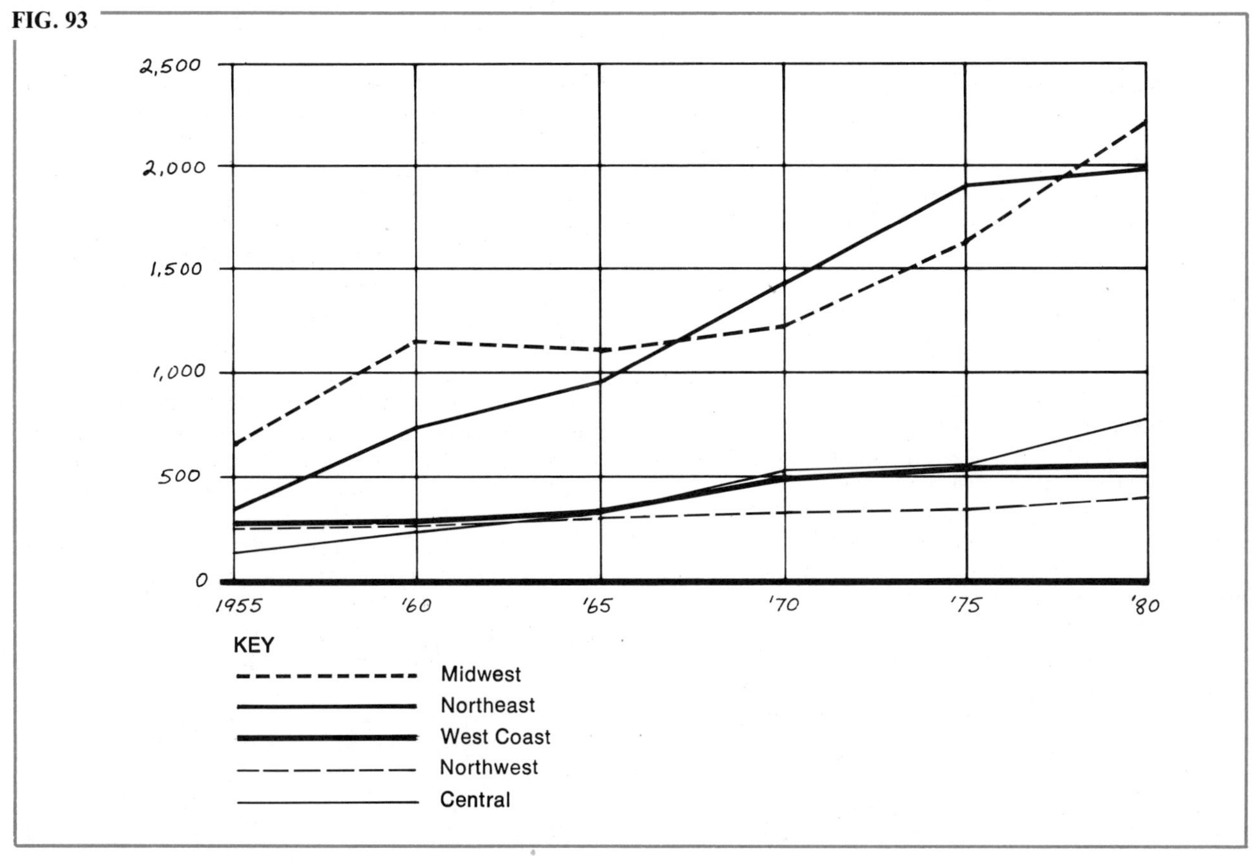
2,500
2,000
1,500
1,000
500
0
1955
'60
'65
'70
'75
'80
KEY
Midwest
Northeast
West Coast
Northwest
Central

This approach helps to some degree. Notice also that it gives the lines a more dramatic movement. The narrower this type of chart, the more you emphasize the line flow. Obviously, the wider the chart (and shallower its height), the less contrast between lines.

Exercise

Prepare this line chart on a sheet of vellum and test it against the answer on page 123.

Industrial Output in Billions of Dollars

	1973	*1974*	*1975*	*1976*	*1977*	*1978*	*1979*	*1980*
Britain	68	76	85	85	92	98	115	135
France	70	82	86	87	86	100	140	152
U.S.	94	110	116	155	165	171	180	194
West Germany	96	112	155	167	170	184	190	165
Italy	130	148	160	125	150	175	189	205
Japan	175	175	180	202	222	232	240	250

Overall dimensions: $4\frac{1}{2}''$ by $2\frac{1}{2}''$
Scale: 0, 50, 100, 150, 200, 250

THE LAYER CHART

This kind of chart requires more work to prepare than the others. There is more arithmetic involved, and it requires a good deal of concentration.

The basic difference in the layer chart is the plotting procedure. Until now each figure you plotted added up from the base 0 line. For example, in Fig. 93 the number of bushels for each territory was plotted for each year along that year's line. This resulted in an immediate comparison of the figures with each other. It does not demonstrate a *total* number of bushels for each year.

This is essentially the function of a layer chart. It shows an accumulation of each figure and their sum total for each year. If you were to convert the line chart in Fig. 93 to a layer chart, it would look like Fig. 95.
The figures plotted in Fig. 95 are precisely the same as in Fig. 93. The difference here is that each figure was added to the next and plotted, resulting in a *stacking* of one plot point over the other. Here is how it is done.

The figure for 1965 starting with the Northeast is 345,605, the first plot point. Then add

FIG. 94

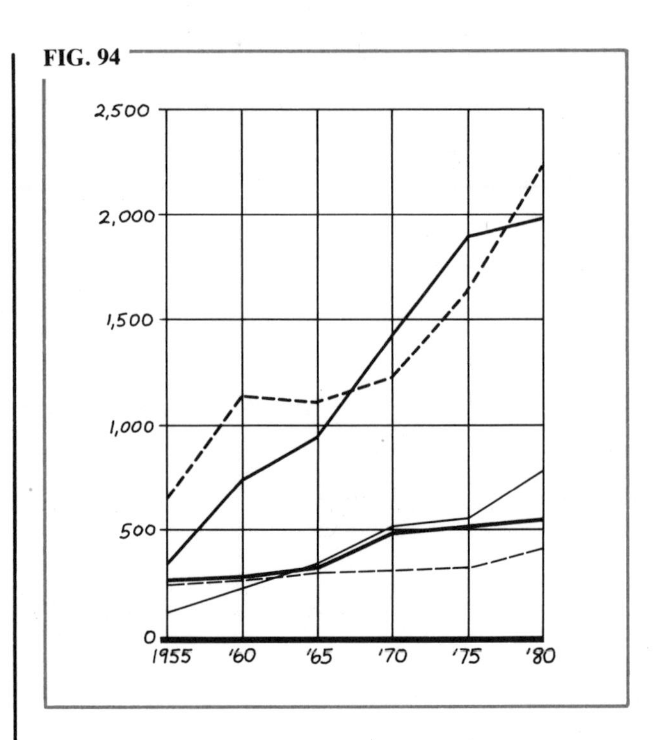

650,700 for the Midwest, which gives 996,305. Plot that point. Now add 255,500 for the West Coast to the 996,305, equalling 1,251,805. Plot that number. Add 250,400 (Northwest) to 1,251,805, equalling 1,502,205, and plot that point. Finally, add 110,100 (Central) to 1,502,205, which gives 1,612,305. Repeat this procedure with each figure for each year and connect the respective points as you proceed. This method of working with the figures is called *cumulating* the figures, i.e., the figures are added successively. One apparent difference in this chart is the overall height of the grid. It must accommodate the highest *totals* of all the figures for *all* categories as opposed to just the single highest figure in each category. This then necessitates a change in scale.

Let's prepare a simple layer chart so that you become more familiar with the procedure.

It will be a comparison of produce costs in actual dollars during a six-month period. The figures to be plotted are the following:

	January	February	March	April	May	June
Meat	\$.77	\$.94	\$1.12	\$1.39	\$1.35	\$1.32
Fruit	.85	.89	.93	1.30	.99	.93
Fish	.67	.83	1.02	1.24	1.18	1.12
Vegetables	.32	.79	.86	1.21	1.10	.94

FIG. 95

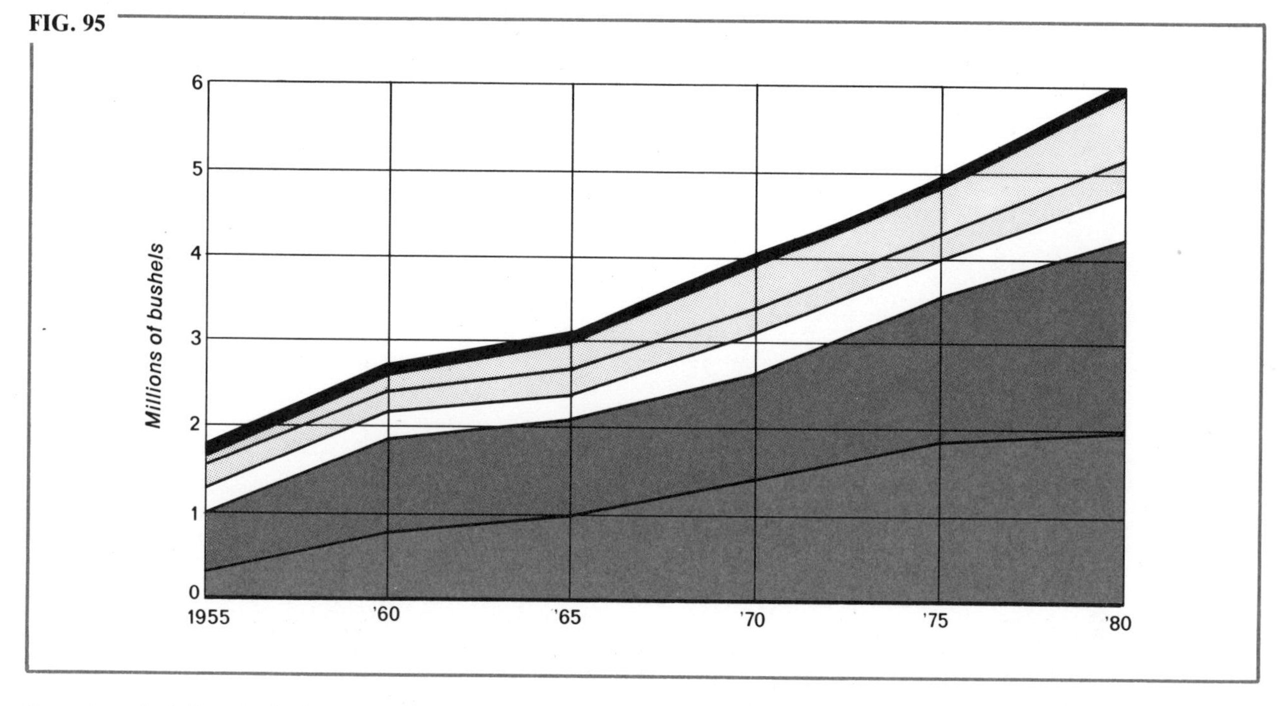

Keep in mind that before you start any kind of chart, you must ascertain the grid and scale breakdown. This is done by taking the largest figure you will be working with and rounding it off to a suitable number. In Fig. 93 the highest plot number is 2,220,400, which is rounded off to 2,500,000. The subdivision is worked out from that highest number.

However, in a layer chart it isn't quite so simple. In order to determine the height of the chart, you must first add the figures for each month so that you can find the highest figure (total) to be plotted, thus determining the highest number on the grid scale.

	January	February	March	April	May	June
Meat	$ .77	$ .94	$1.12	$1.39	$1.35	$1.32
Fruit	.85	.89	.93	1.30	.99	.93
Fish	.67	.83	1.02	1.24	1.18	1.12
Vegetables	.32	.79	.86	1.21	1.10	.94
	$2.61	$3.45	$3.93	$5.14	$4.62	$4.31

The largest total, which is for April, is $5.14; thus the height of the chart can be $5.25 (or $5.50 if you prefer). As a matter of fact, the height of the chart can be any denomination you choose. It could conceivably be $10.00. But that may give the chart a peculiar appearance, and why waste all that space? Unless, of course, you are trying to create a particular effect. The height could just as easily be $5.14. It is not uncommon to use the highest plot point as the height of the chart. We will review the approach to this in another chapter.

Generally, the highest figure in a chart is the next rounded off figure after the highest plot point. This makes the chart easier to lay out and gives the most effective shapes from the figures. It is really a matter of personal preference.

Let's use $5.50 as our top figure. Your scale can be in quantities of $0.50, giving you a total of 11 segments. Your grid scale for plotting can be divided into 5 equal parts, each representing $0.10.

To save time, copy the grid in Fig. 96 or trace it onto a piece of vellum.

You have a choice of two approaches to follow when plotting a layer chart. One is to cumulate each *vertical* column and connect as you go along from column to column (month to month). For example, cumulate and plot all the figures up the January column, then the February column, and so on.

The other method is to plot one *category* at a time *horizontally* for each month, cumulating one complete category at a time. For example, plot all

FIG. 96

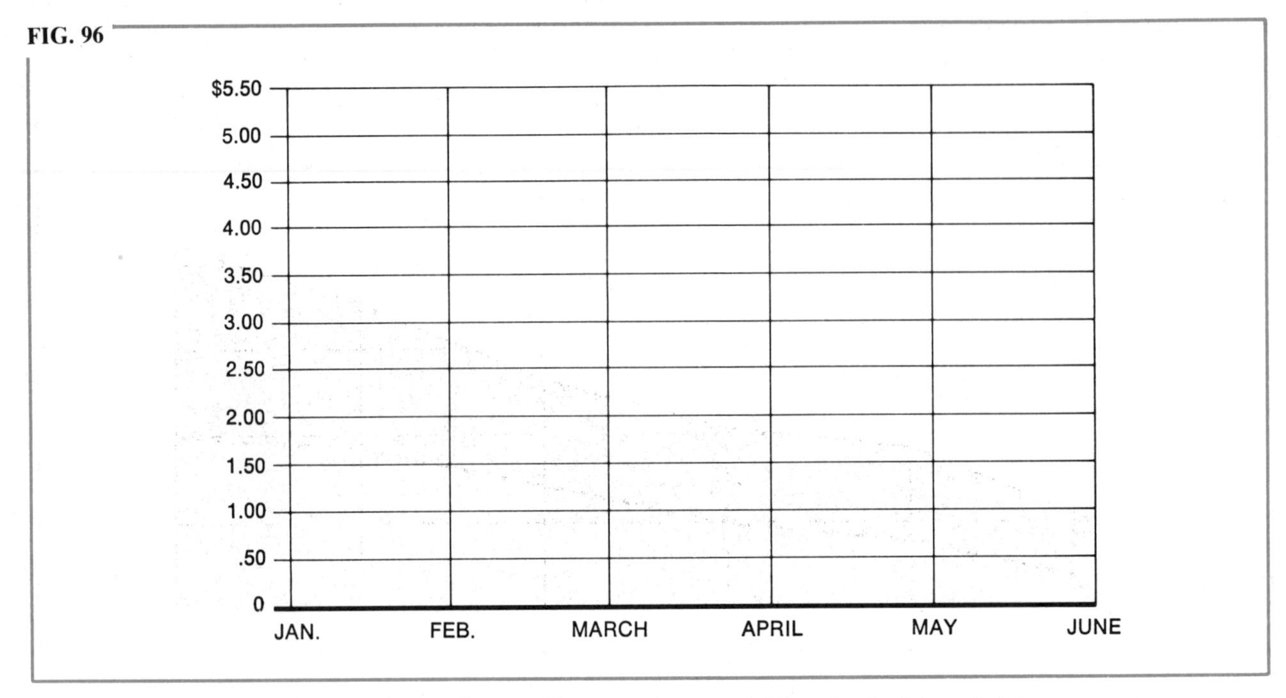

the figures for vegetables (one for each month), then all the figures for fish (one for each month), adding each as you plot from month to month.

Either method is suitable.

Let's try it using the first method. Follow these steps: (Fig. 97)

FIG. 97

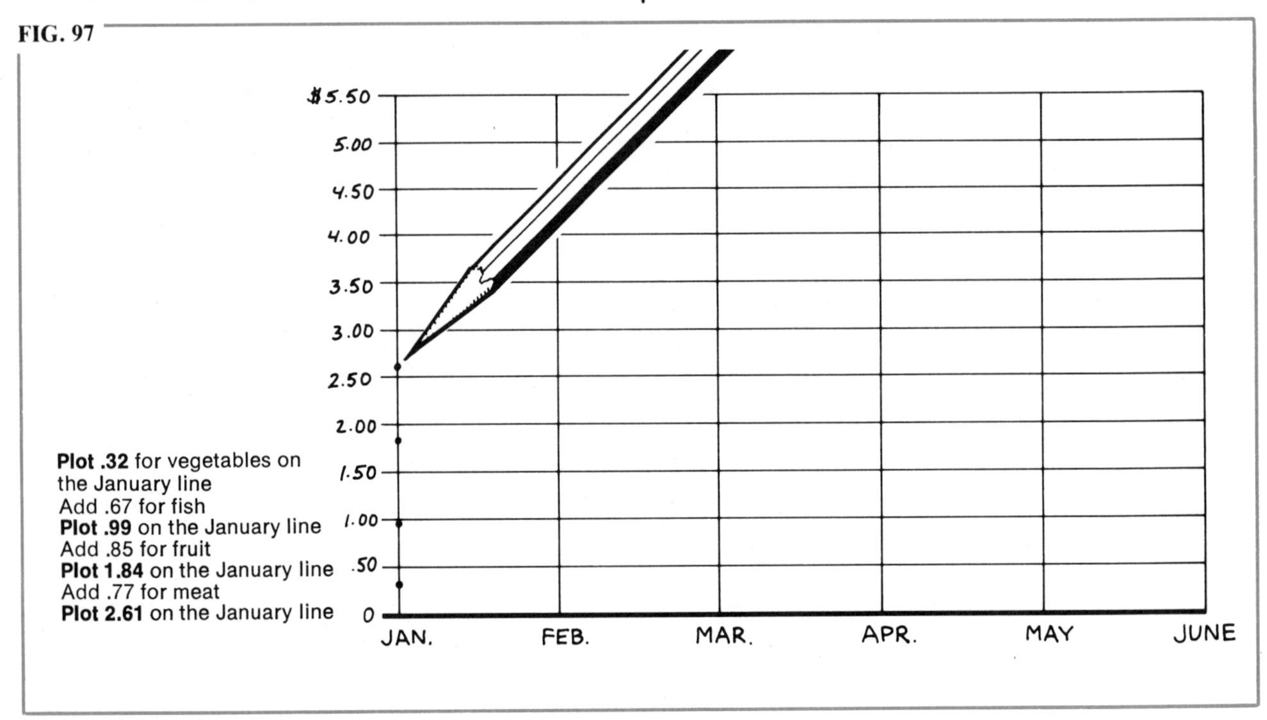

One advantage of adding one figure to the other is that it is much simpler to plot their totals than to plot the actual figures. It is easier to find .99 on the chart than it is to plot .32 and then attempt to find the plot point for .67 starting at the .32 point.

Now for the February line: (Fig. 98)

FIG. 98

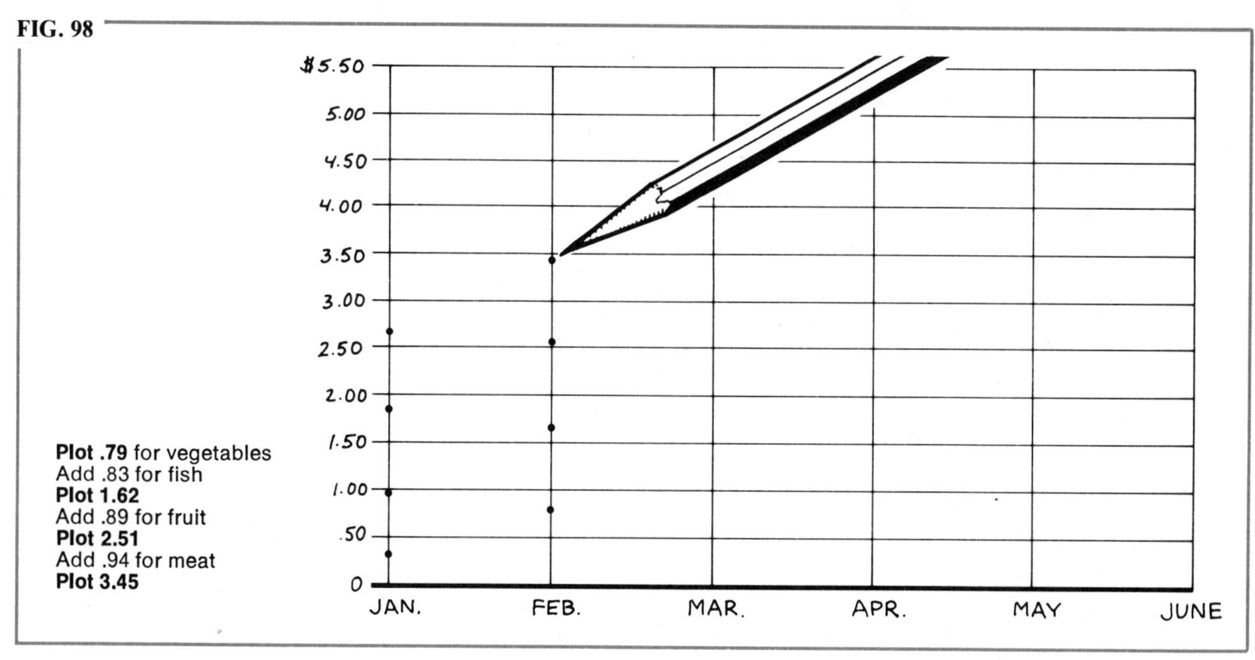

Repeat this procedure for March: (Fig. 99)

FIG. 99

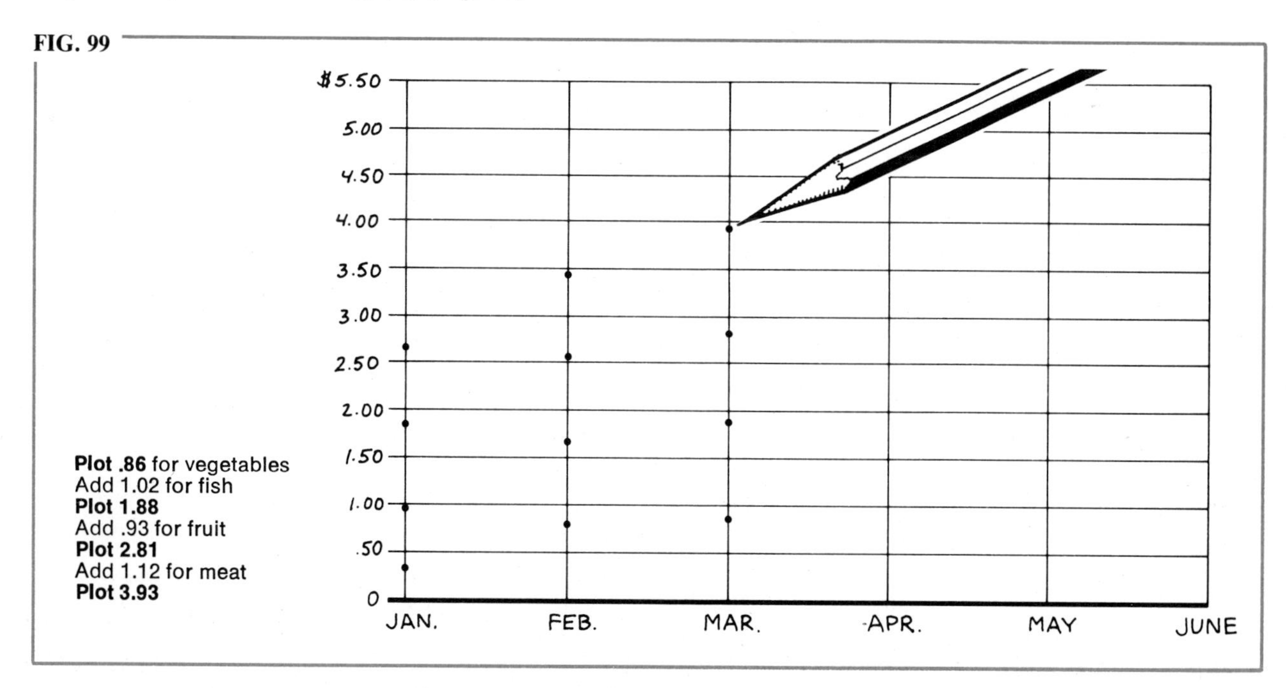

April: (Fig. 100)

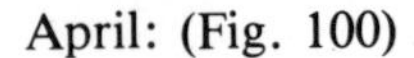

FIG. 100

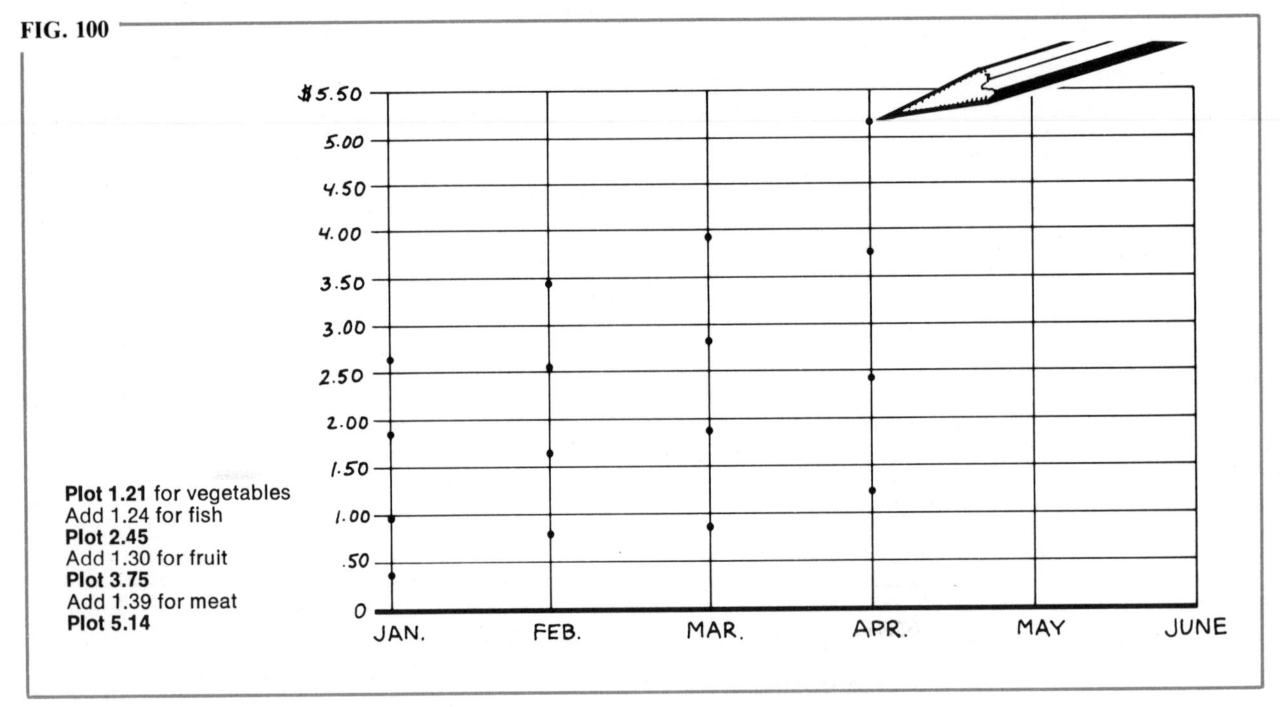

May: (Fig. 101)

FIG. 101

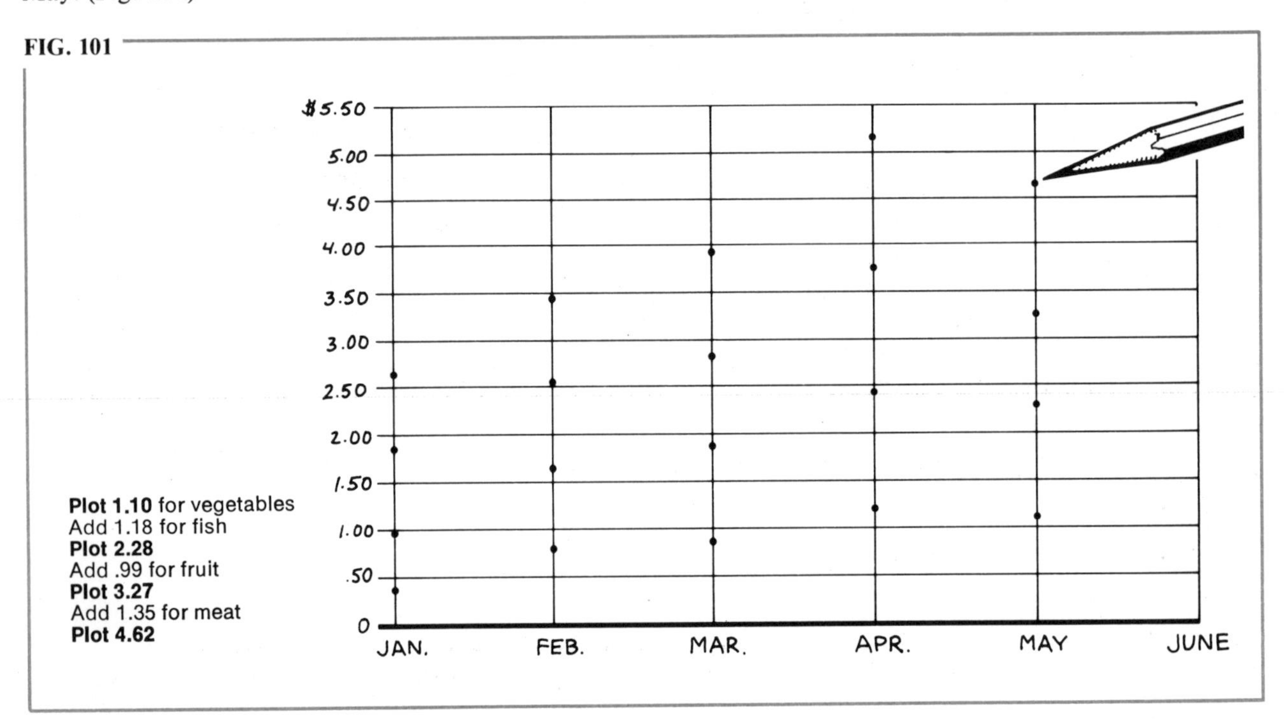

June: (Fig. 102)

FIG. 102

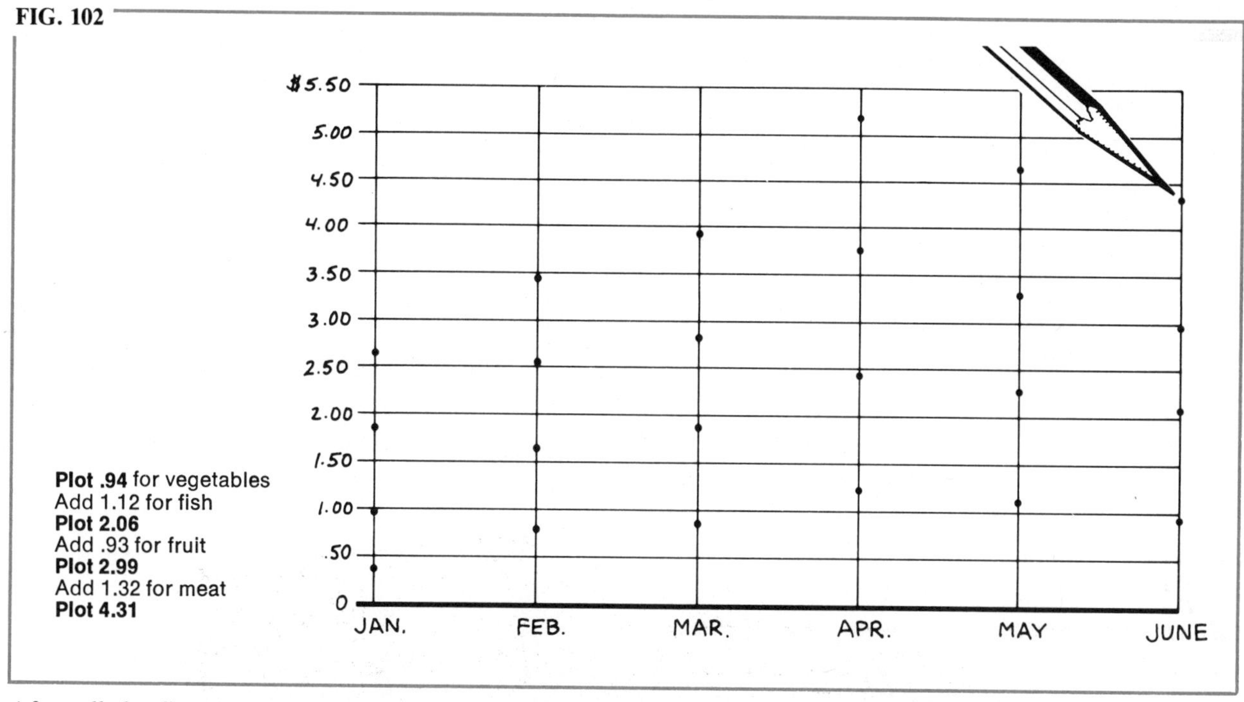

After all the figures have been plotted and connected, the layers become apparent (Fig. 103).

FIG. 103

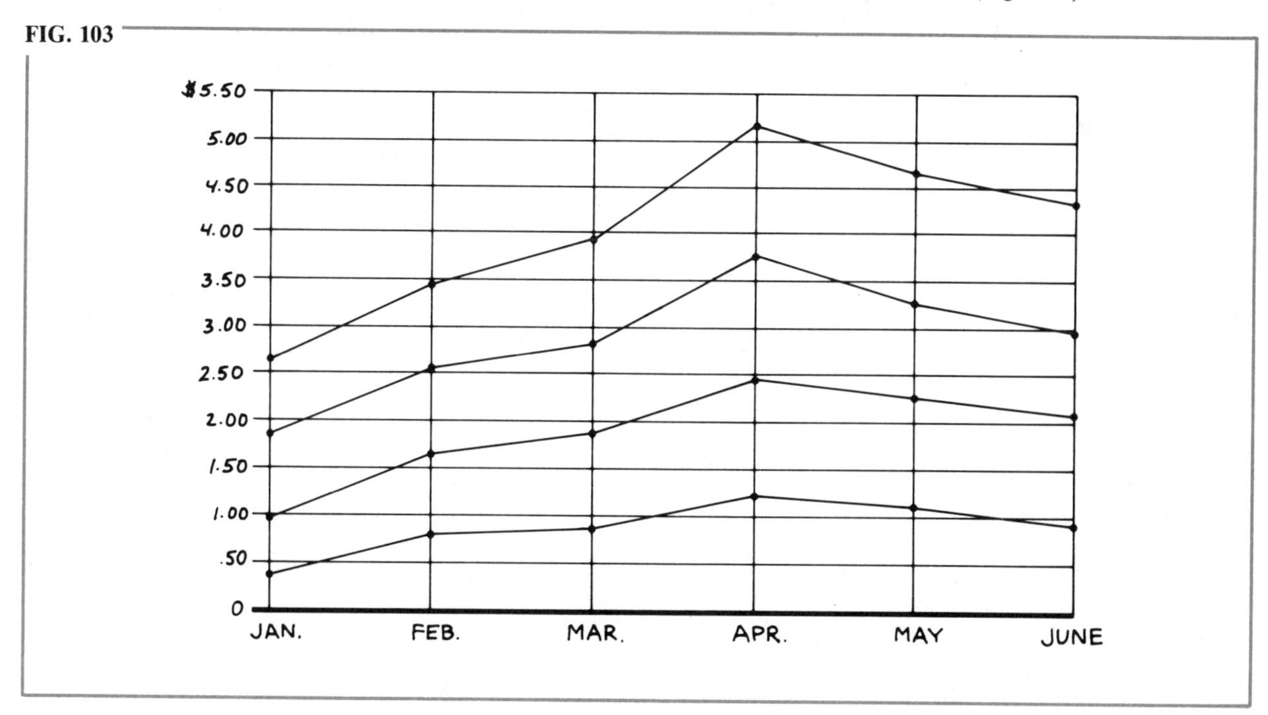

The top line is frequently made heavier than the others to stress the total. Of course, each layer is to be filled in, usually each with a different color, texture, or value. It's all a matter of style (Fig. 104).

This type of chart is always impressive and quite informative. It shows the progressive variations of the individual category over a period of time. Simultaneously it compares the progression of each category, illustrating its contribution to the whole. And it also dramatizes the variations of totals of all the categories over a given period.

FIG. 104

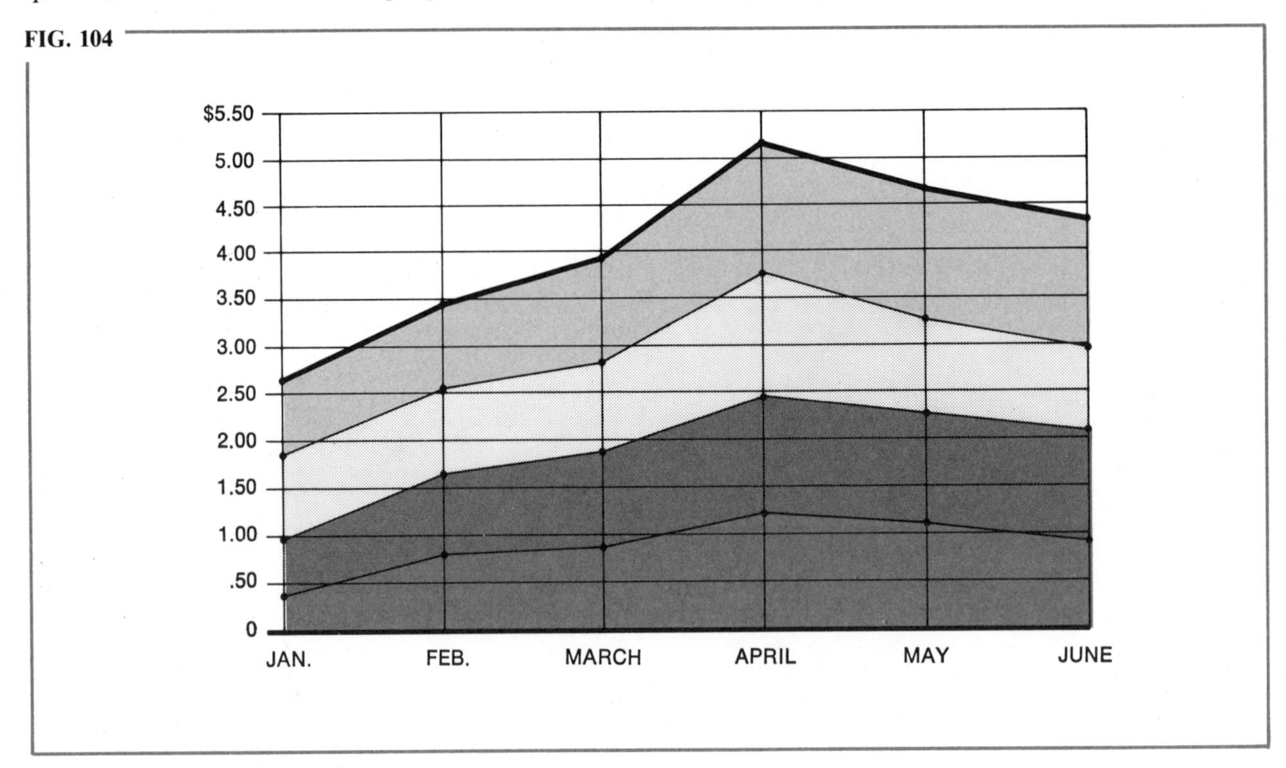

A LAYER CHART VARIATION—100% LAYER CHART

The chart in Fig. 103 deals with *specific figures* in each category. Its purpose is to show and compare how much each has contributed to the whole, in addition to illustrating the monthly variations in the totals. The *percentage* layer chart, on the other hand, is not concerned with illustrating the various totals in each period of time. Nor is it interested in the *actual figures* of each unit of comparison, but rather in the relationship of each to their *total,* i.e., not specifically how much each category contributed, but actually what its percentage of the total is.

Look at January's total ($2.61) in Fig. 104. The dollar value of each item within the total actually represents a percentage of the total figure, which in itself is 100%. The same holds true for February and for each month in the chart, the sum being 100%, and each category within each month is a percentage of 100%.

Let's work out some percentages.

The procedure for determining the percentage is to divide the *total into* the number being converted. For example:

What percentage of 2.61 is .32? Simply divide 2.61 into .32 to obtain 12%. Do the same for .67; divide 2.61 into .67 to obtain 26%. The next figure is .85; divide 2.61 into .85 to obtain 32.5%. What percentage of 2.61 is .77? When you add them all, the total should be 100%.

What exactly are we doing here? We are converting figures into percentages. 2.61 becomes 100%, and each of the figures that add up to the

2.61 are likewise converted into percentages. When added, they total 100%.

Now, if each month's figures total 100%, then the top of the chart will not vary at all. On the other hand, the categories within each month will vary according to their percentages.

Let's begin. It requires a fair amount of simple arithmetic. Here are the figures to be converted into percentages:

	January	February	March	April	May	June
Meat	$.77	$.94	$1.12	$1.39	$1.35	$1.32
Fruit	.85	.89	.93	1.30	.99	.93
Fish	.67	.83	1.02	1.24	1.18	1.12
Vegetables	.32	.79	.86	1.21	1.10	.94
	$2.61	$3.45	$3.93	$5.14	$4.62	$4.31

To save time, here are the converted percentages:

	January	February	March	April	May	June
Meat	29.5%	27%	28%	27%	29%	31%
Fruit	32.5	26	24	25	21	21
Fish	26	24	26	24	26	26
Vegetables	12	23	22	24	24	22
	100%	100%	100%	100%	100%	100%

After converting to percentages you must then cumulate them for plotting. The figures to be plotted are indicated with a check mark.

January		February		March	
	12✓		23✓		22✓
	+26		+24		+26
	38✓		47✓		48✓
	+32.5		+26		+24
	70.5✓		73✓		72✓
April		**May**		**June**	
	24✓		24✓		22✓
	+24		+26		+26
	48✓		50✓		48✓
	+25		+21		+21
	73✓		71✓		69✓

The overall dimensions of the chart will be the same as in Fig. 96: $4\frac{5}{16}''$ by $2\frac{11}{16}''$. The monthly vertical divisions will remain the same, but the horizontal grid will have to change.

The top line of the chart represents 100%. We can divide the scale into any number of parts we find convenient. Let's use 10.

Do you remember how to do this? Why not stop reading at this point and see if you can do it on your own?

FIG. 105

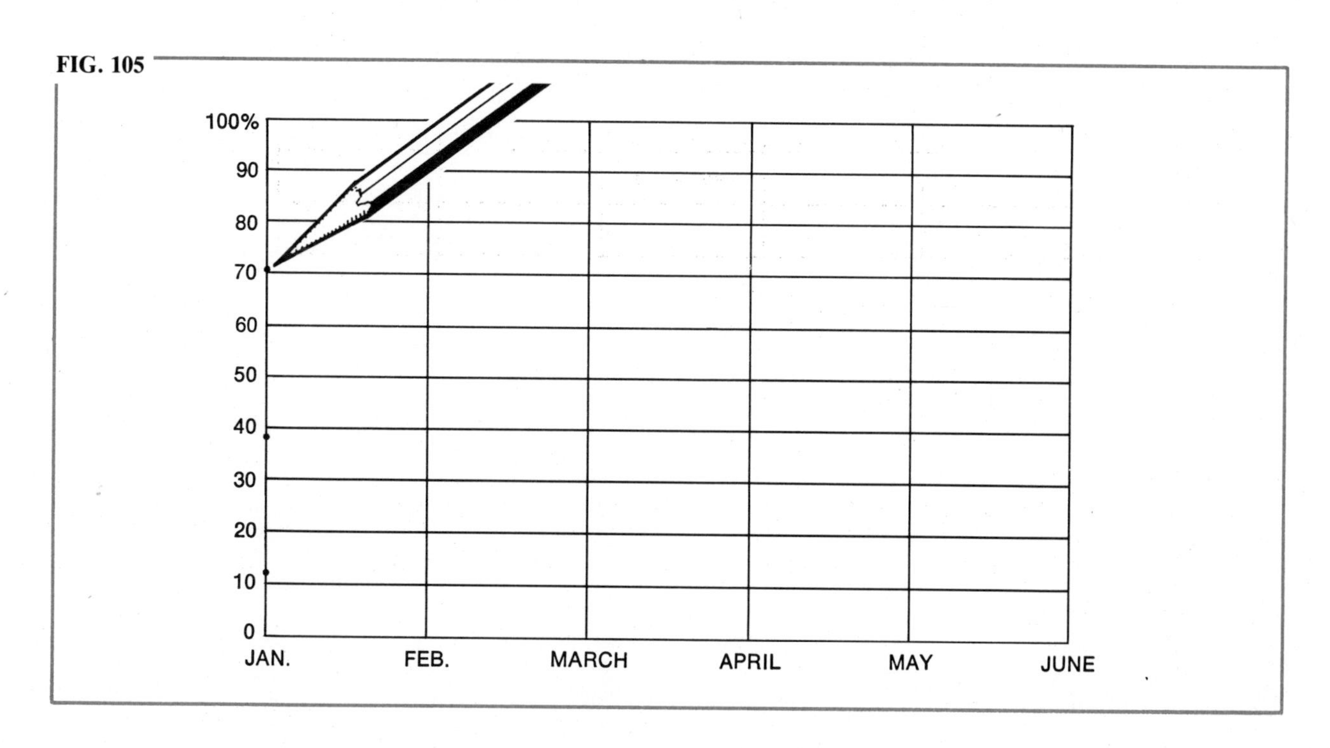

To subdivide this grid into 10 equal parts, let's use a pica ruler for a change. Angle the 20 pica mark between the base line and the top line, and tick off every 2 picas.

Prepare your grid scale for one segment. Try dividing it into 10 parts. It will be very tight, but it can be done—very carefully. You are ready to plot.

At this point you may be vulnerable to error. When plotting one often gets confused while translating the plot figure in terms of the grid. Make it a habit to study your grid scale for a moment before you begin plotting. Be sure you have clearly in your mind precisely what the grid represents. This grid scale represents 10% and is broken into 10 segments, each one is 1%. Now start plotting.

The first percentage is 12. Align your scale between 10 and 20 and tick off the second mark.

Next is 38. Move your scale to the 30–40 space. Tick off the eighth mark.

The next figure is 70.5. Place your scale between 70 and 80. You will have to do this one by eye. Just a hair over 70. That's the end of that month. There is no need to plot 29.5; the remainder of the line to 100% is 29.5 (Fig. 105).

Plot the rest of the figures. When you are through plotting, connect all the plot points, and that's it! (Fig. 106).

Note the difference between the two types of charts. Each tells its own story rather effectively.

FIG. 106

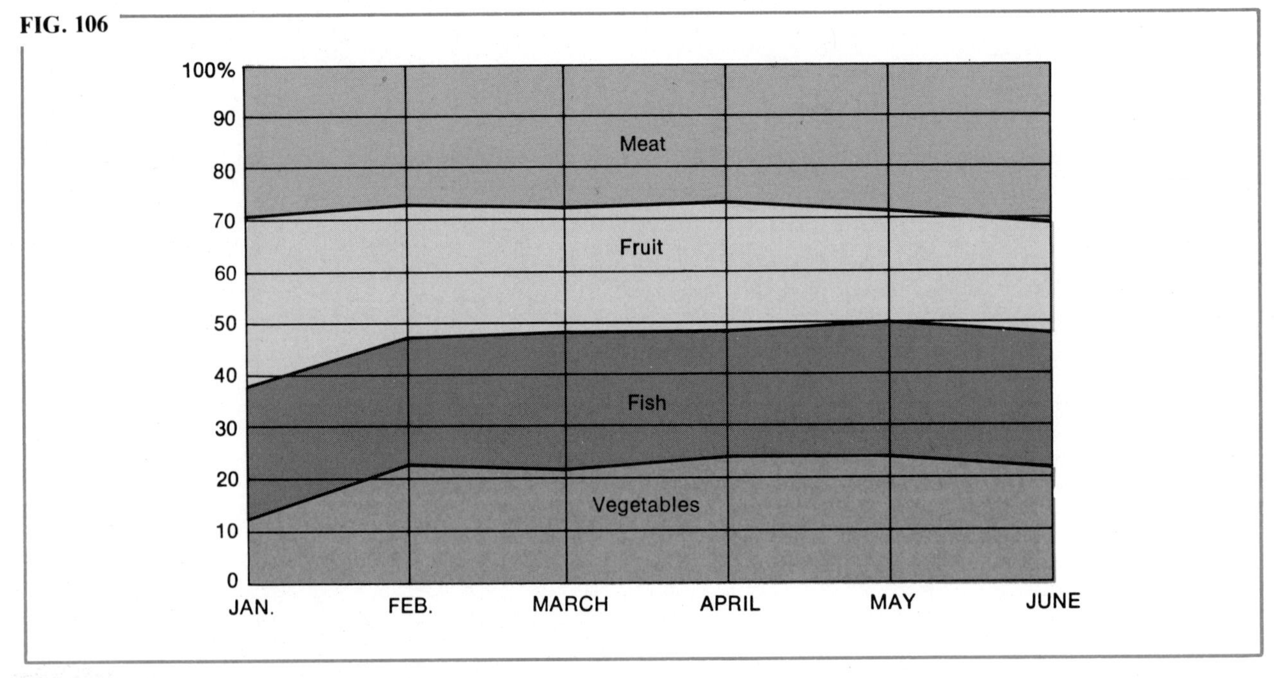

Exercise: Prepare a 100% layer chart.
Overall dimensions: $6\frac{7}{8}''$ wide by $5\frac{1}{4}''$ high
Scale: 0 through 100 in 5% increments

Answer on page 124.

Percentage of Disposable Family Income

	Housing	Food	Clothing	Transportation	Medical Care	Recreation	Miscellaneous	Savings
1950	24%	28%	10%	15%	12%	4%	5%	2 %
1955	29	26	12	13	11	2	4	3
1960	22	25	18	19	9	4	2	1
1965	23	22	17	10	12	4	6	6
1970	24	21	17	19	7	3	5	4
1975	28	28	14	19	8	1	1	1
1980	29	27	7	18	10	4	2	3

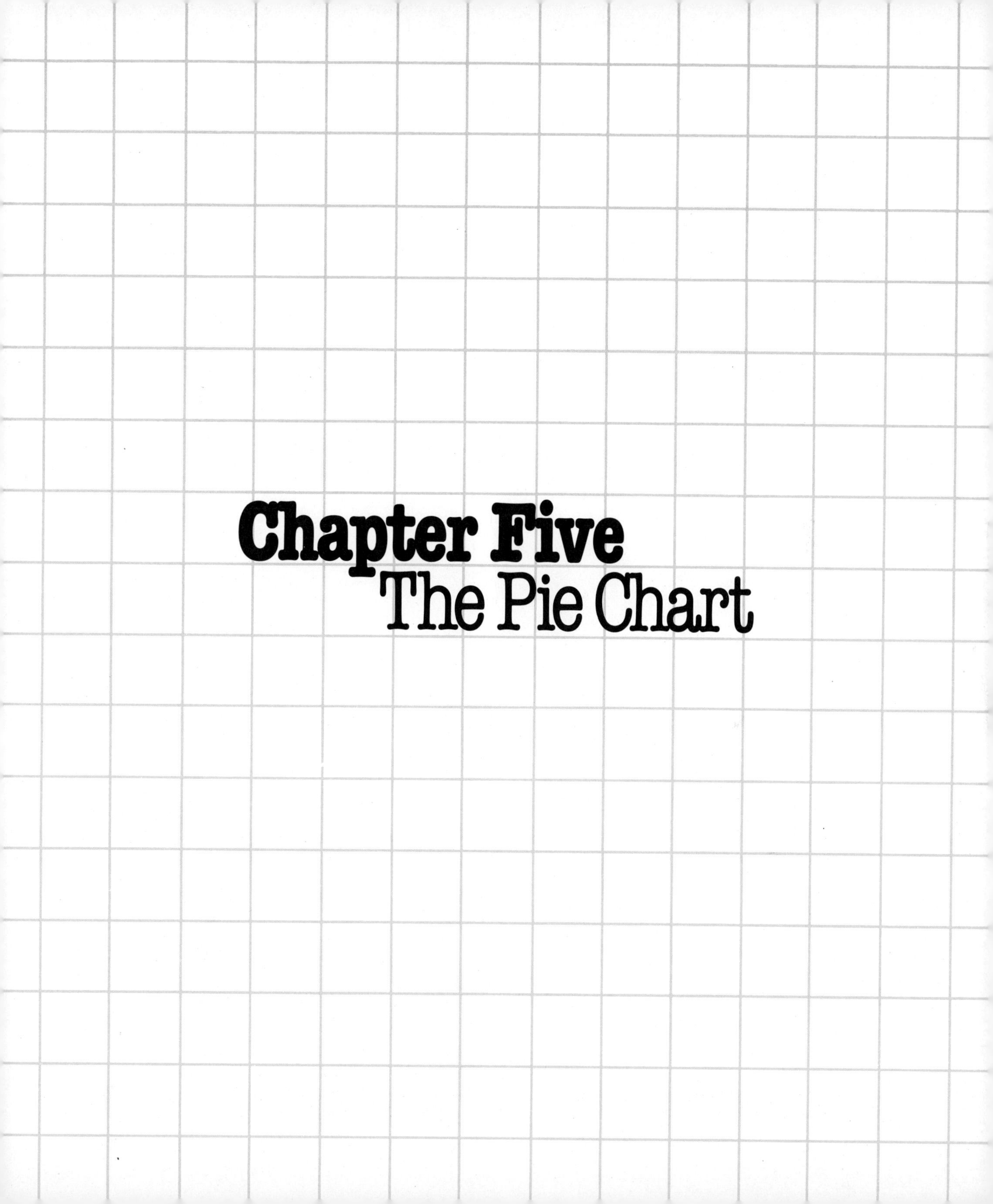

Chapter Five
The Pie Chart

In principle the scale for a pie chart is quite similar to that for a bar or column chart. You need to determine a segment of the overall chart and break the segment down into subdivisions with which you can plot your figures. However, creating a scale for a pie chart is somewhat different from doing so for a bar chart. For example, in a bar chart your scale represents a portion of the bar in a straight line. In contrast, a scale for a pie chart represents a portion of the circle in a curved line. Furthermore, you can easily utilize principles of geometric projection with a bar chart, whereas these principles are not altogether applicable to a circle chart. To the best of my knowledge there is no known *simple* geometric approach to this problem.

Nevertheless, the least complicated and most adequate method of preparing a pie chart scale is to use an adjustable triangle (Fig. 107).

FIG. 107

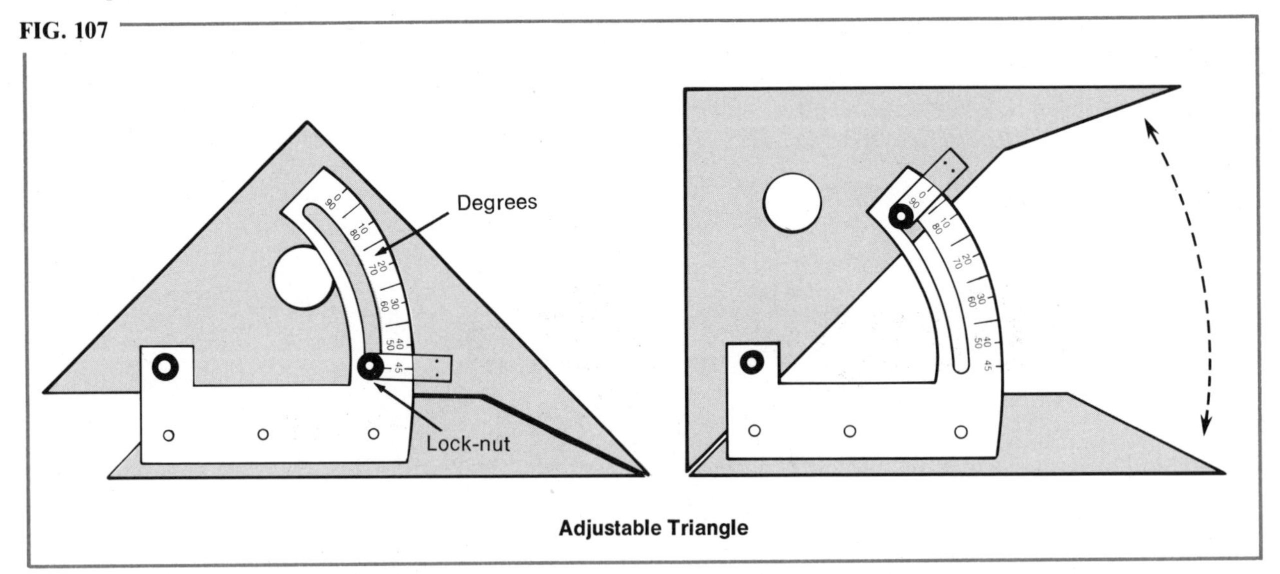

Adjustable Triangle

The curved scale in the center represents 45°. With this tool you can divide a circle into as many as 360 equal parts (if necessary).

A pie chart generally represents a 100% total of two or more items. For example, a magazine readership study may reveal that 54% of all its readers are women aged 32–37; 27% are women of age 25–32; 16% are 18–25 years of age; and 3% are younger than 18. The sum of these figures is 100%, which completes the pie. These figures are graphically illustrated in a pie chart in Fig. 108.

FIG. 108

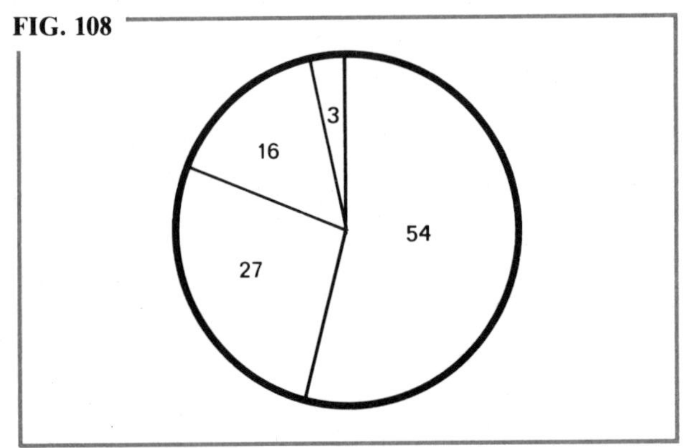

In order to plot portions of a circle, you must first subdivide it into a given number of equal parts. Of course, 100 parts would be ideal but unnecessary. All you really need to do is construct a wedge (pie) of the circle that represents $\frac{1}{10}$ of its circumference. Then divide that segment into 10 equal parts and you'll have the scale to plot with.

You know that the circumference of a circle is an arc of 360°. Therefore, when it is divided into 10 equal parts, each part is 36° (360 ÷ 10 = 36). Now further divide one of these segments into 10 equal parts (36 ÷ 10 = 3.6). This segment (3.6) is actually $\frac{1}{100}$ of the circle (360 ÷ 3.6 = 100). What you now have is a scale that can easily be used to find percentages of the total circumference of a given circle. But rather than talk about it, let's prepare such a scale by first dividing a circle into 10 equal parts.

On a piece of vellum construct a circle 3″ in diameter (Fig. 109).

FIG. 109

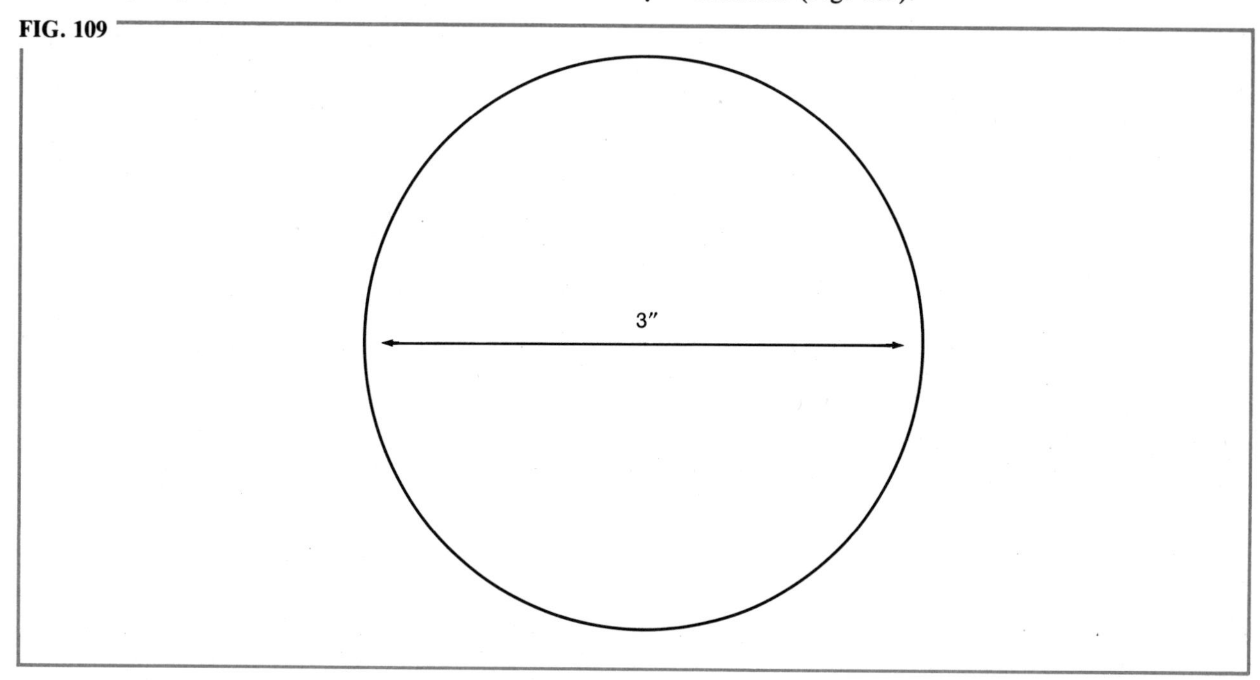

With a T-square and triangle draw a vertical line through the center (A), dividing the circle in half.

FIG. 110

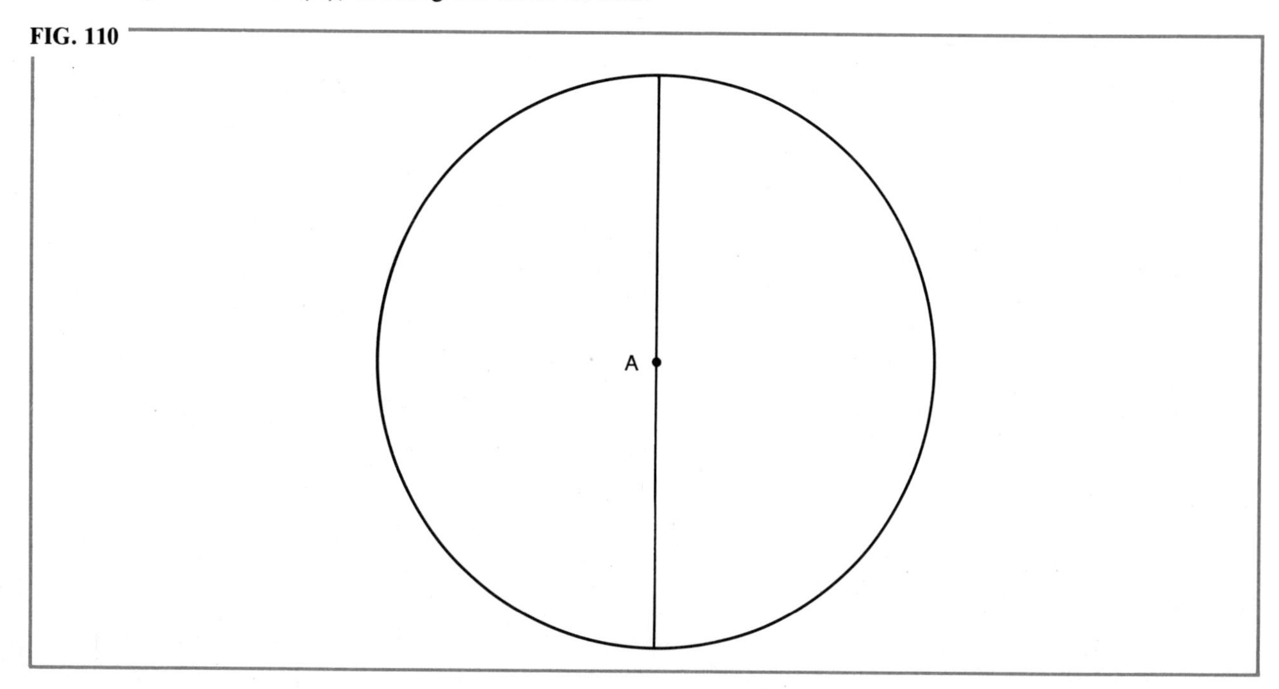

Set and lock your adjustable triangle at 36° (Fig. 111).

FIG. 111

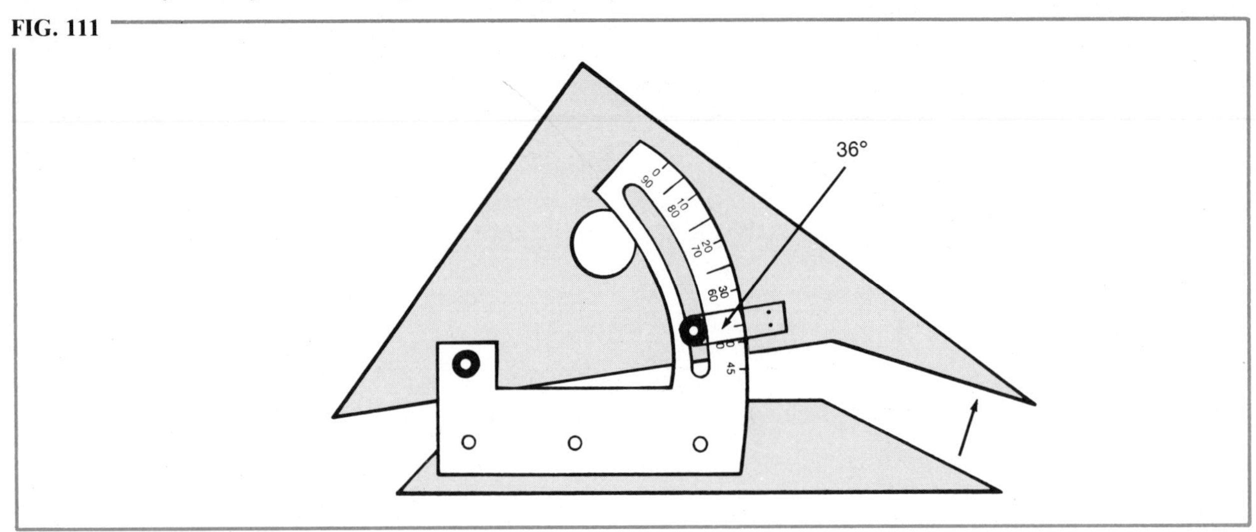

Position your triangle on a T-square and draw a line along the 36° angle through A to the circumference at both ends (Fig. 112).

FIG. 112

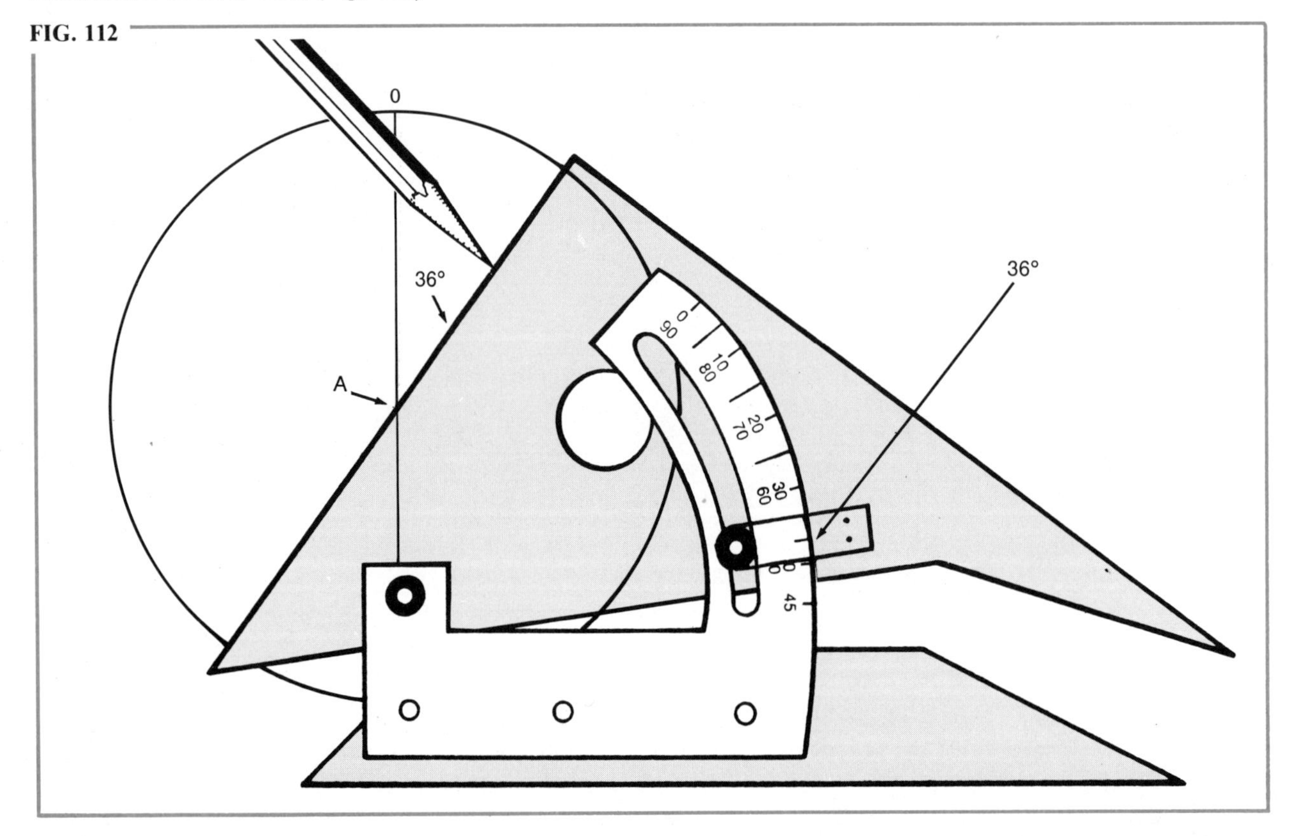

Now open the triangle an additional 36° to 72°, lock it, and construct another line at the angle through center A (Fig. 113).

FIG. 113

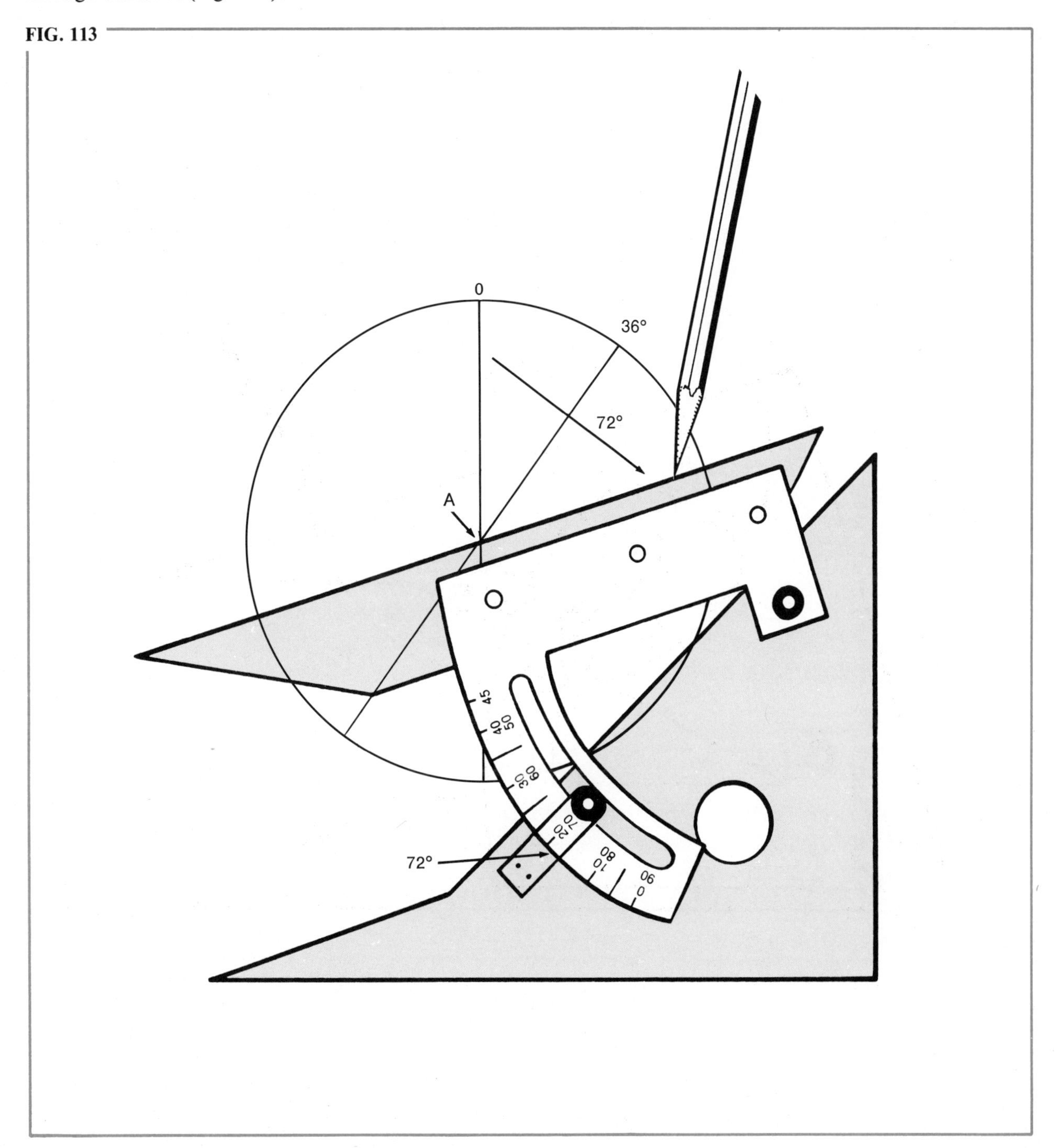

Next ($72^\circ + 36^\circ$) is 108°. Notice that the setting for 108° is the same as for 72°, but the triangle must be positioned the opposite way (Fig. 114).

FIG. 114

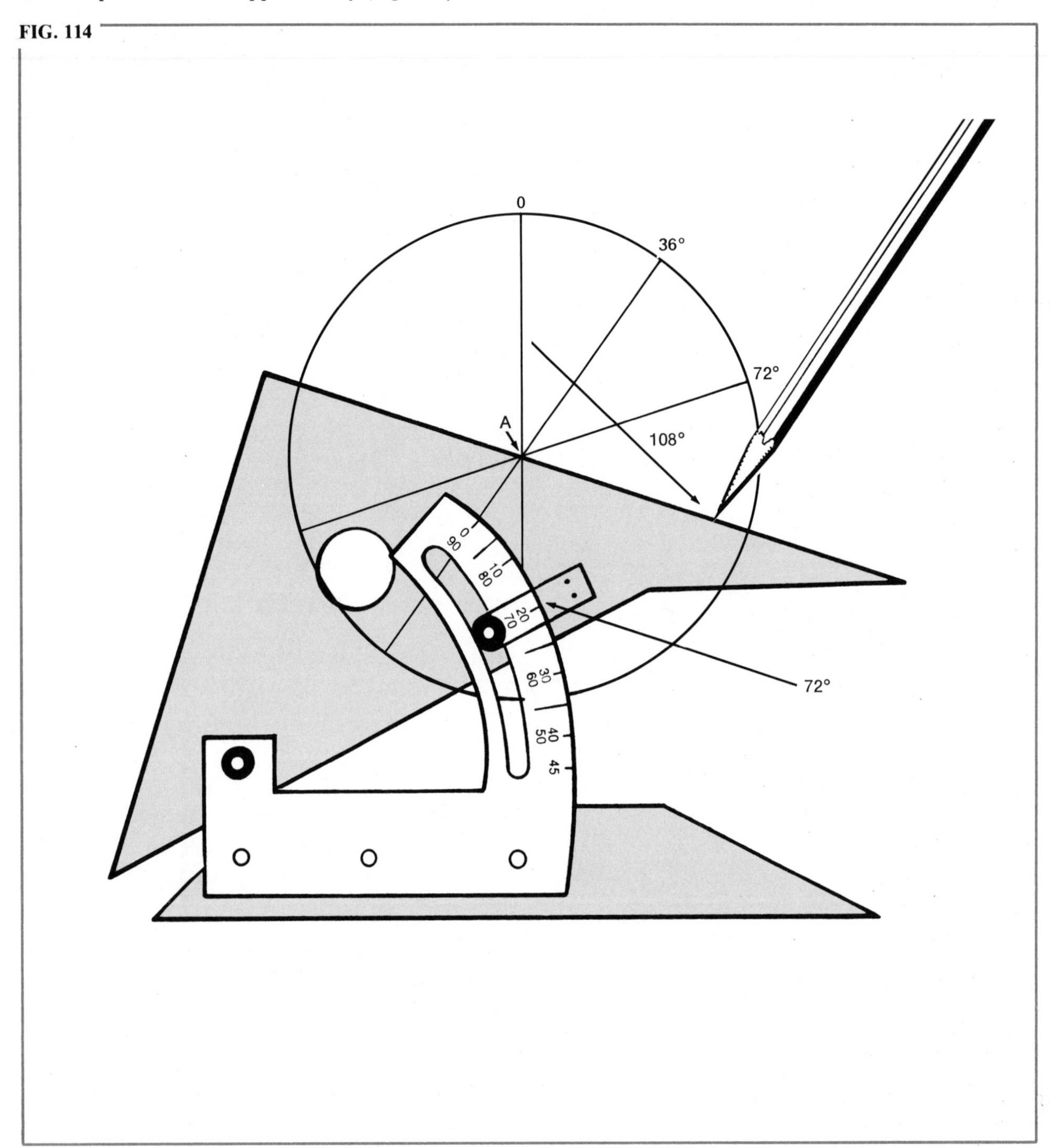

Finally, draw your last line at 144°, which is the same setting as 36° with the triangle positioned differently. You have now divided a circle into 10 equal parts. It's really easy as "pie"(Fig. 115).

FIG. 115

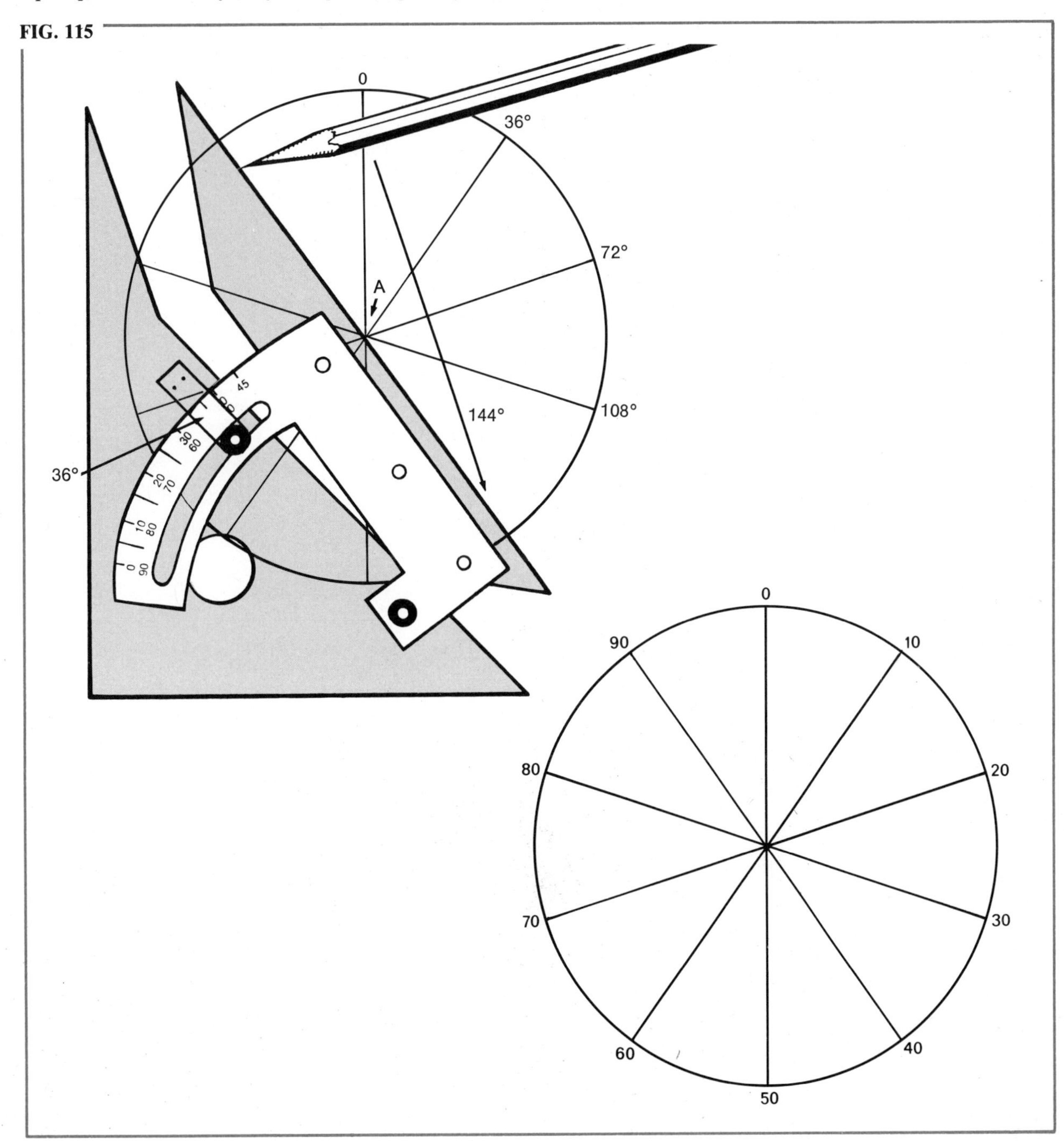

The next step is to divide one of these 10 segments into 10 equal parts by drawing a line every 3.6°. This will be a lot easier to do if we add the 3.6 figures in advance, obtaining 3.6, 7.2, 10.8, 14.4, 18.0, 21.6, 25.2, 28.8, and 32.4, as in Fig. 116. If you were to continue in this manner around the entire circle, you would have divided it into 100 equal parts, but this is not really necessary (Fig. 117).

FIG. 116

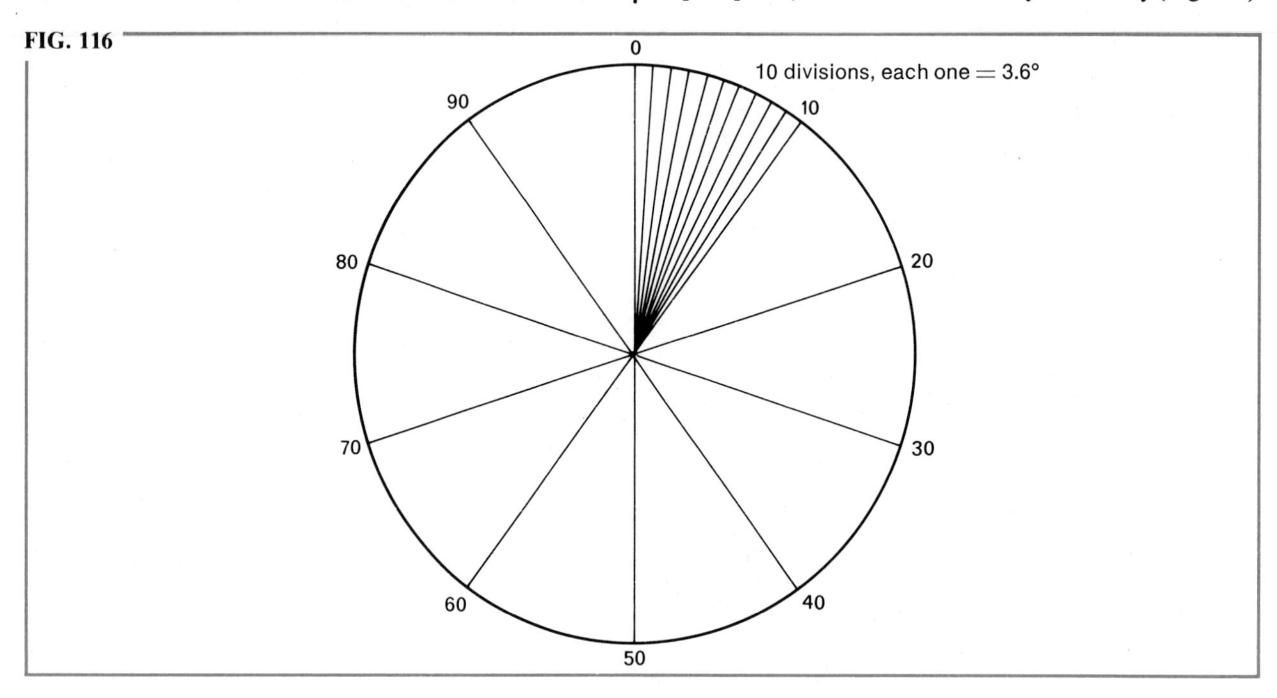

FIG. 117

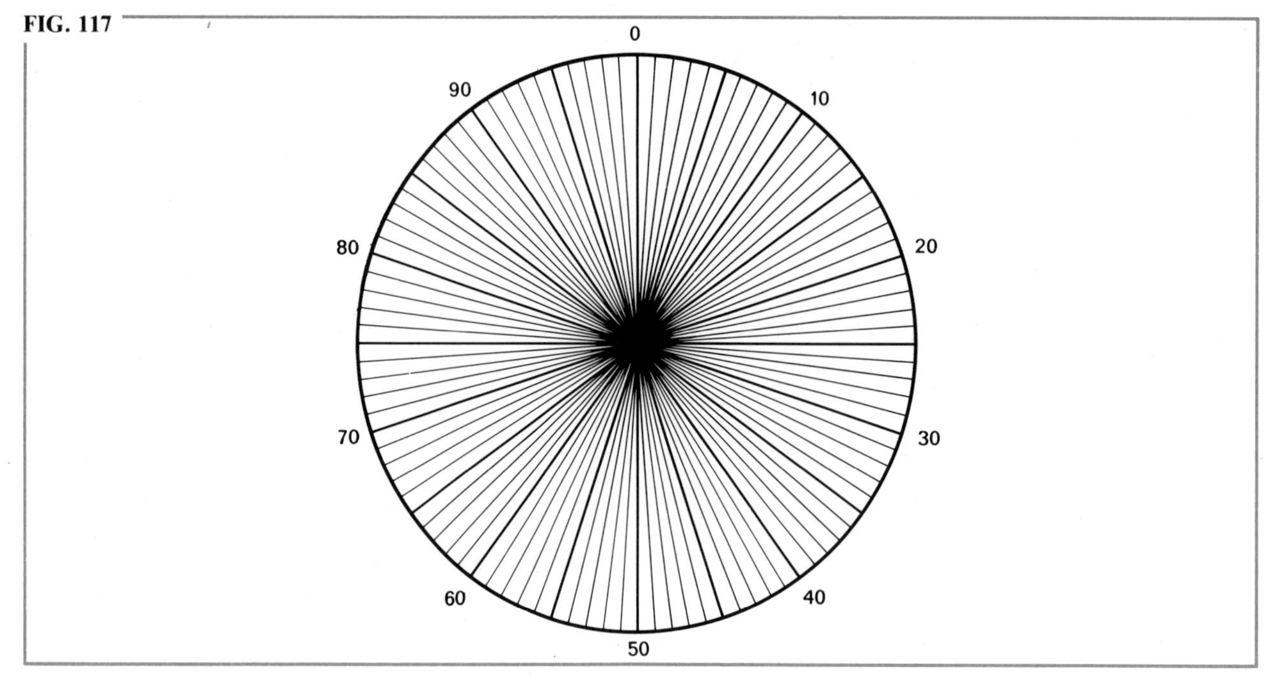

All you actually need for plotting is the $\frac{1}{10}$ scale (Fig. 116). Here is how it can be used to plot the figures shown in Fig. 108. Take the scale prepared in Fig. 116 and construct a square around it about $\frac{1}{2}''$ from the circumference. Extend all the segment lines beyond this square (Fig. 118). Then trim the whole thing down by cutting along the square around the circle. Label the segments.

FIG. 118

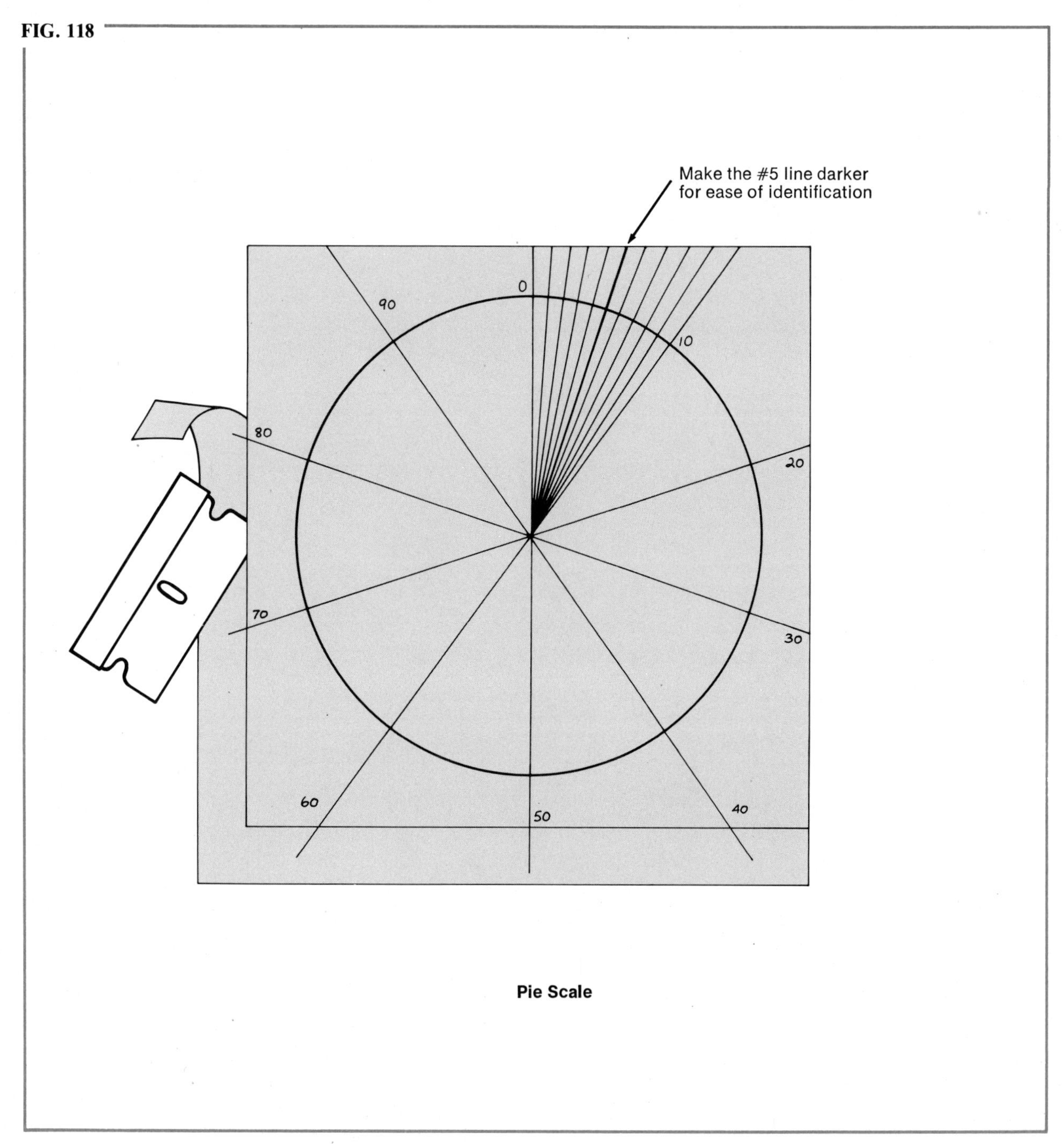

Pie Scale

On a sheet of white bond construct a circle 2″ in diameter and draw a vertical line from the center point A to the circumference 0 (Fig. 119).

Next cumulate the figures to be plotted: 54, 54 + 28 = *81*, 81 + 16 = *97*, 97 + 3 = *100*.

Place your scale directly over the constructed circle, aligning both center points (A) and the lines A–0 (Fig. 120).

With your needle-pointed 6H pencil, tick off the 50, 80, and 90 marks. Remove the scale and lightly connect the center point A with the 50 mark (it will extend beyond the circumference). Repeat this with both the 80 and 90 marks (Fig. 121).

Reposition the scale so that its A–0 line is directly over the A–50 line of the art (Fig. 122).

FIG. 119

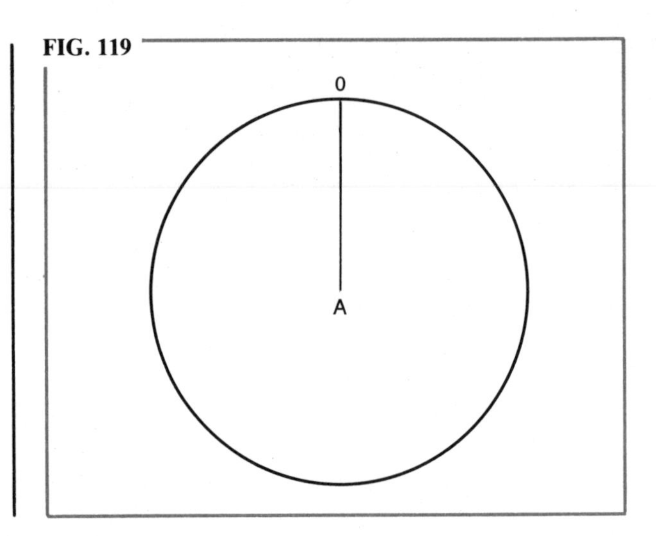

FIG. 120

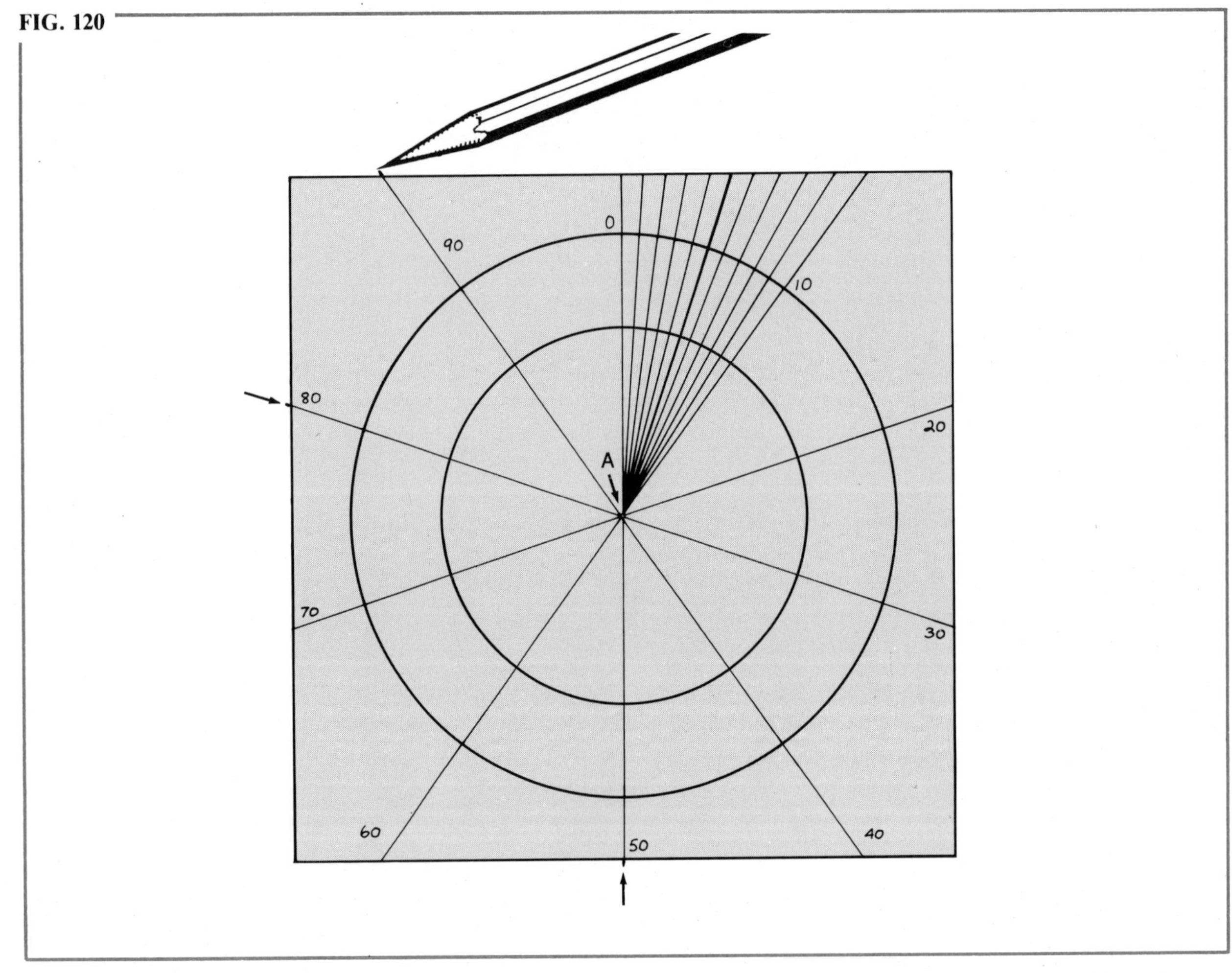

FIG. 121

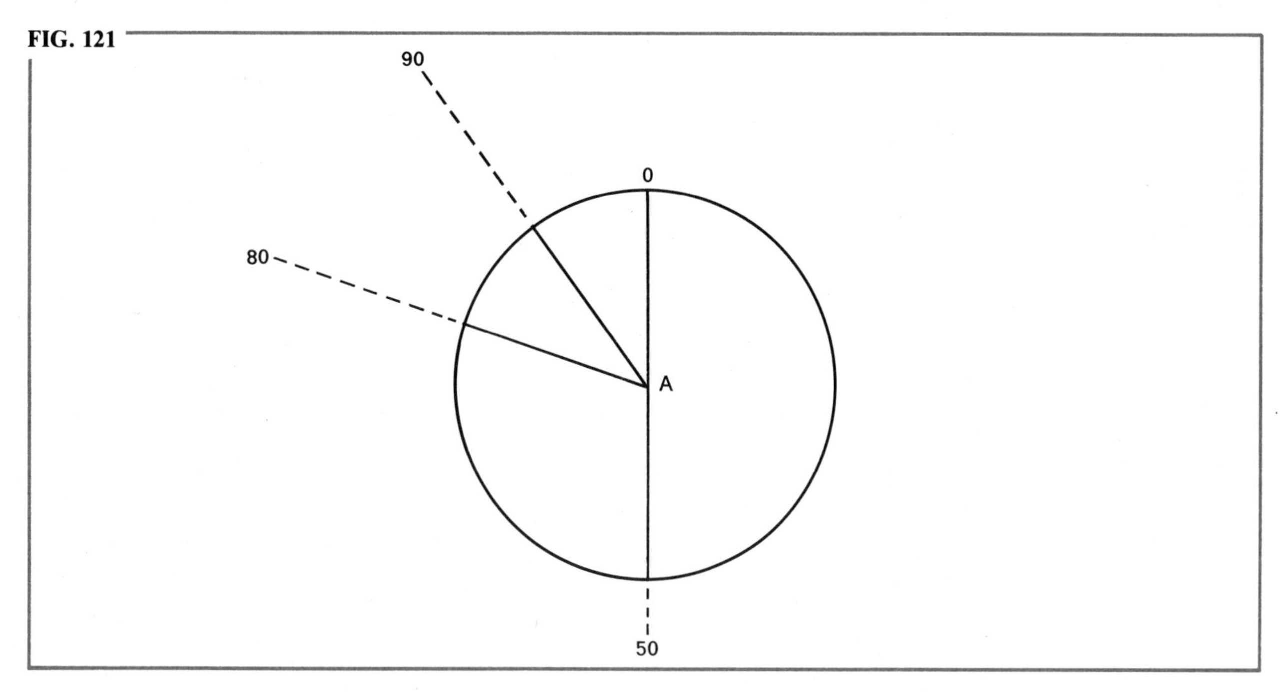

FIG. 122

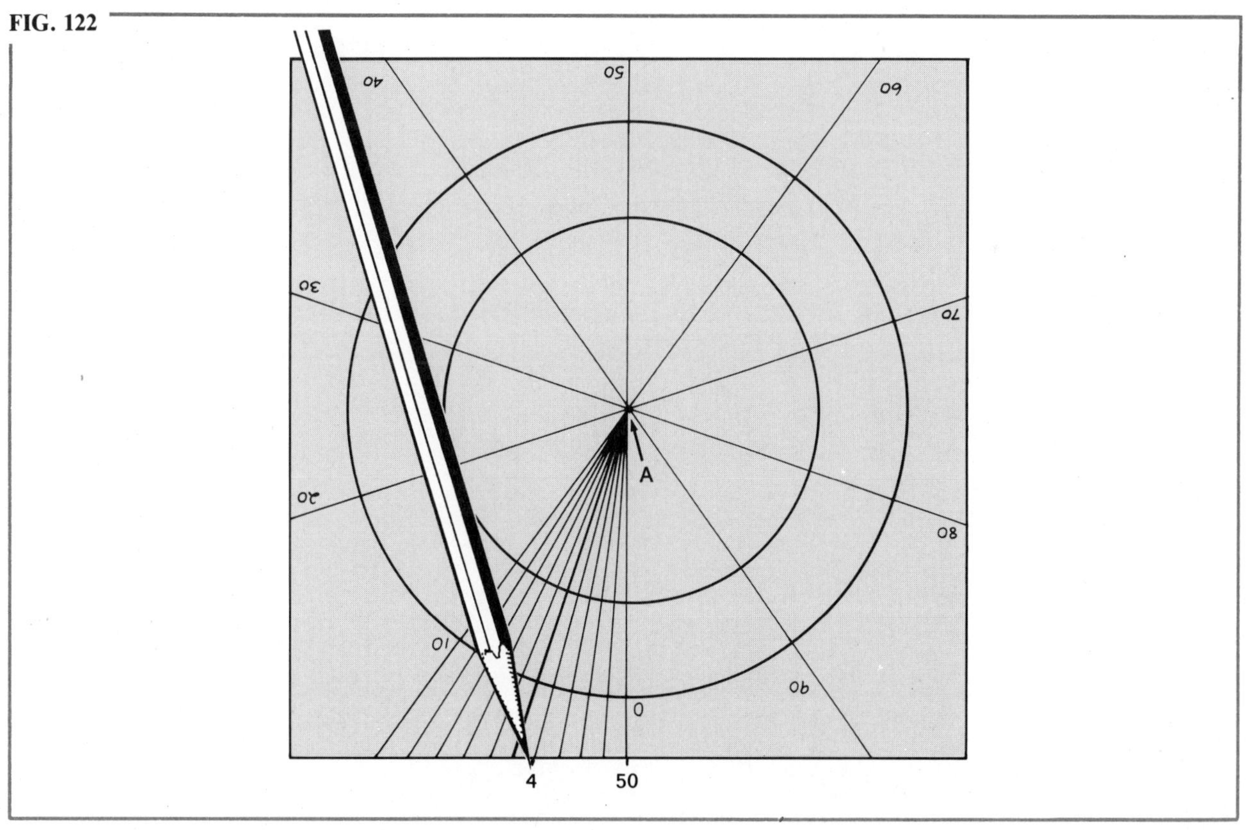

Count off four segment lines and place a mark off the edge of the scale. Remove the scale and connect A and 4. This is the first pie segment, 54 (Fig. 123).

Once again, reposition the scale so that A–0 aligns with the A–80 line. Count off one segment line and place a mark off the edge of the scale (Fig. 124).

FIG. 123

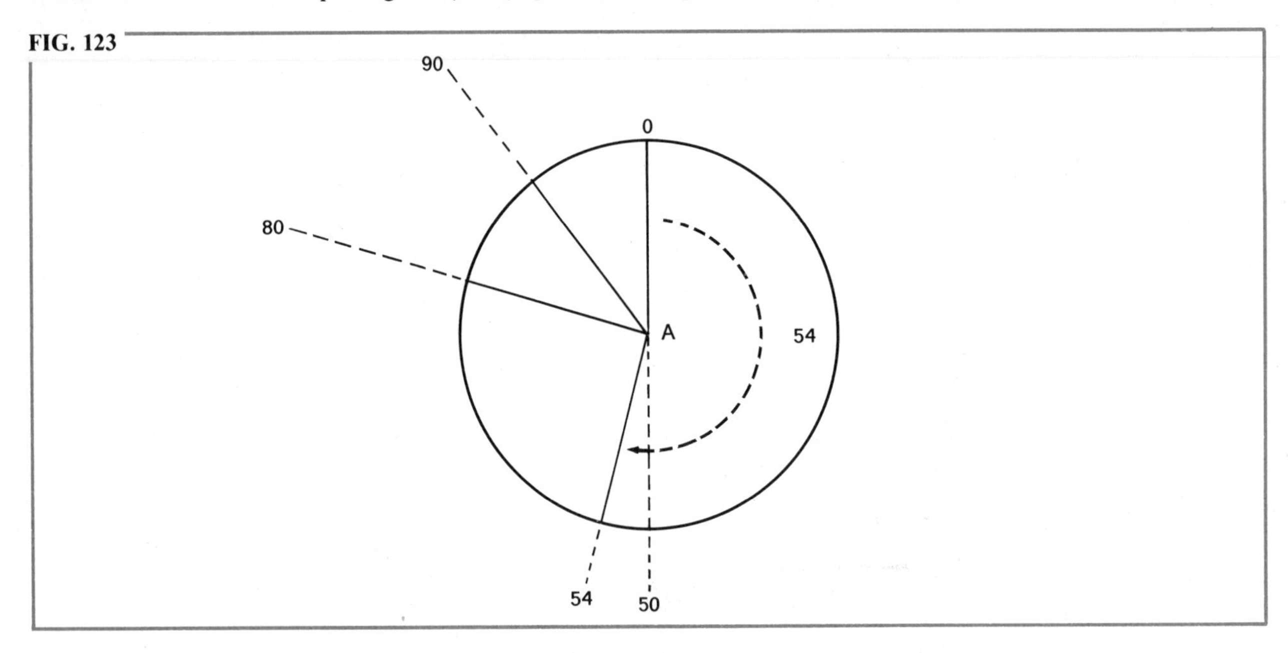

FIG. 124

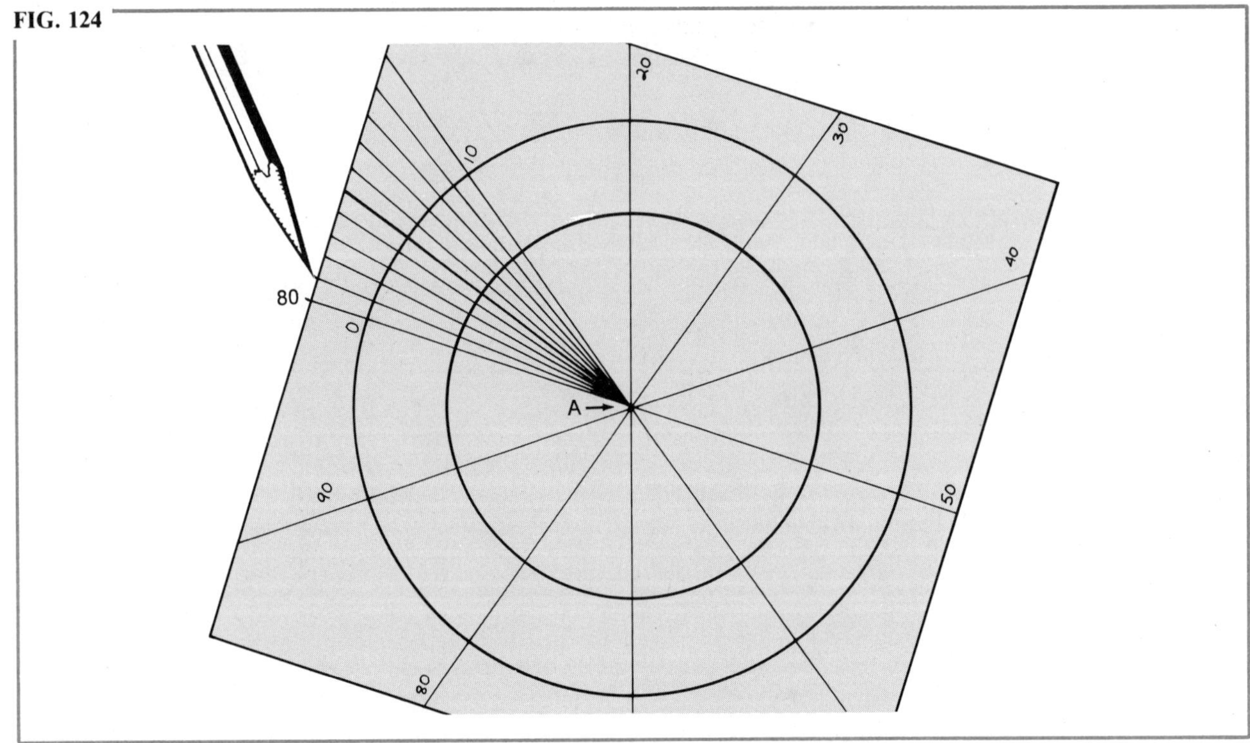

Connect A and 1. This is the second pie segment, which is actually 27 (Fig. 125).

Position the scale's A–0 line over the A–90 line. Mark off the seventh segment (Fig. 126).

FIG. 125

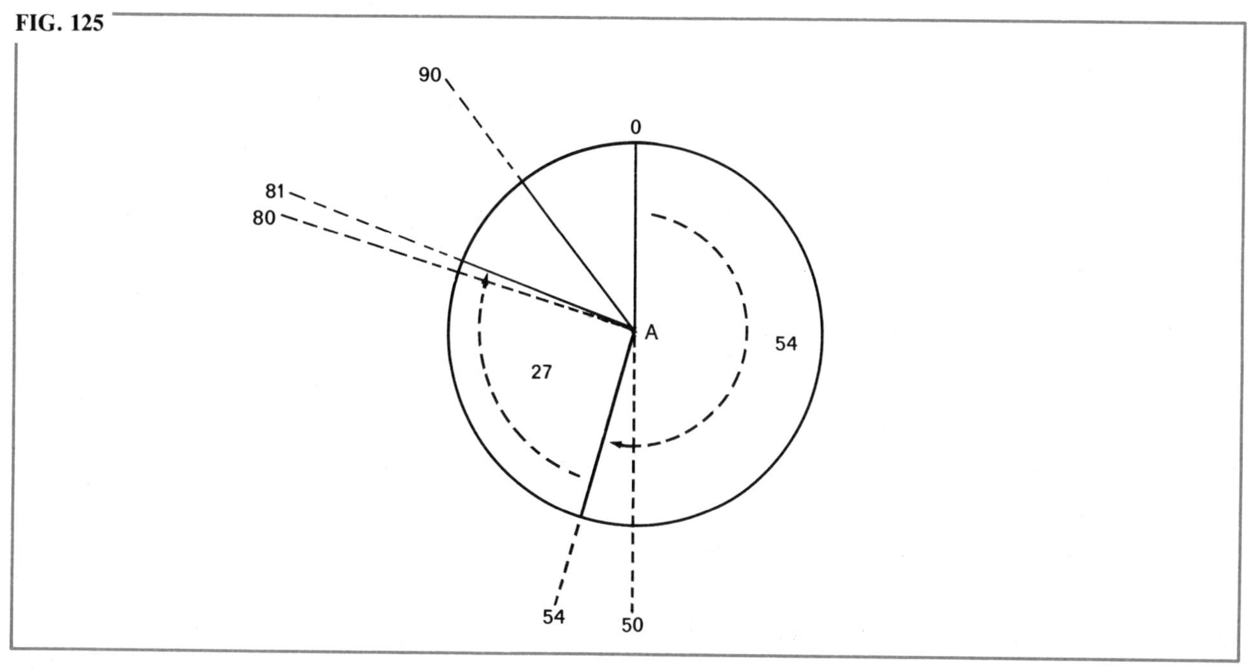

FIG. 126

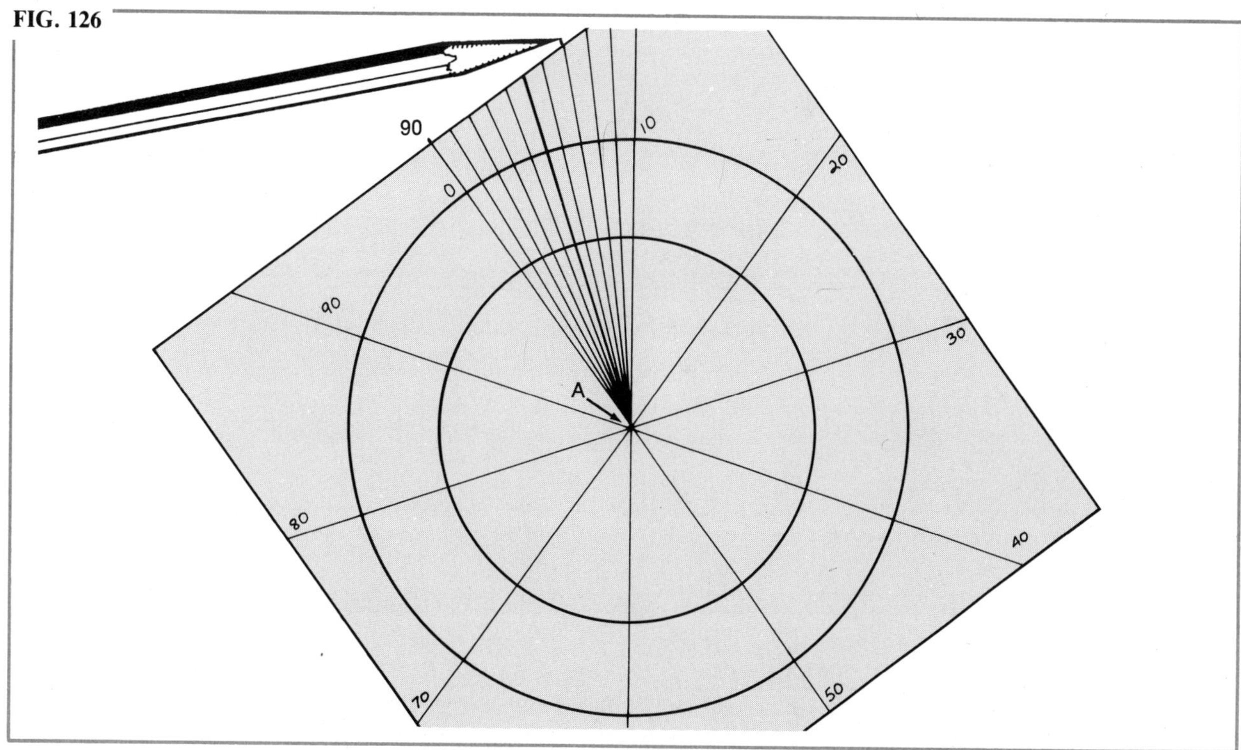

FIG. 127

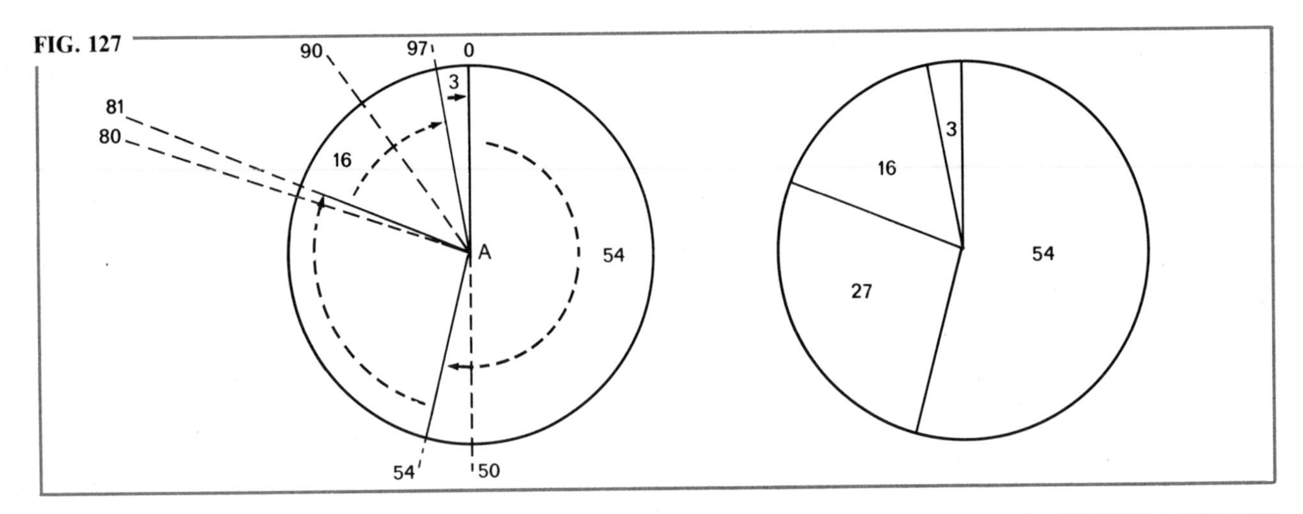

FIG. 128

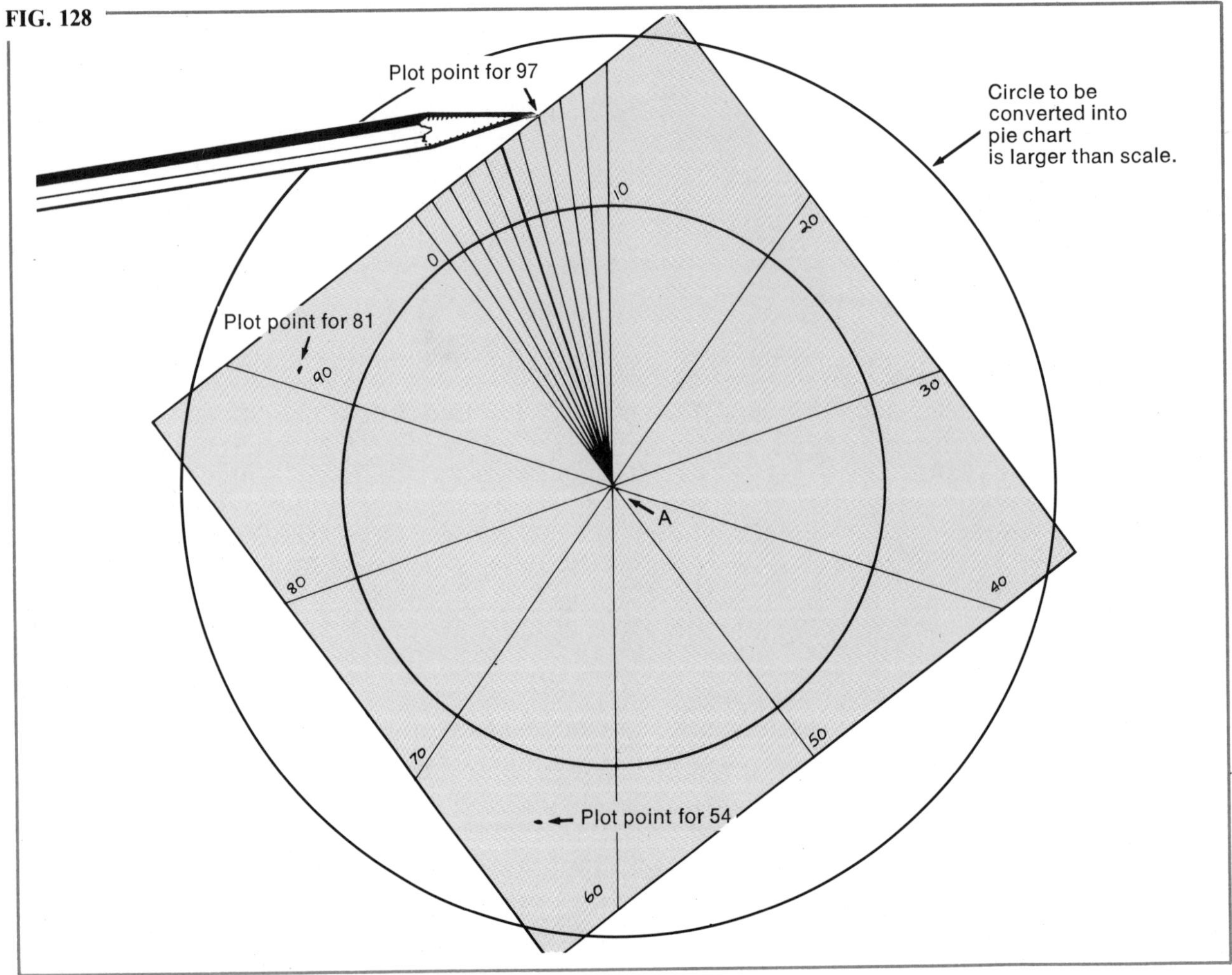

Connect A and 7. This is the third pie section, which is actually 16. Because it is the last line plotted, the remaining pie section is the last percentage figure, 3. Your pie chart is complete (Fig. 127).

No doubt you have by now observed that your scale need not have the same radius as the circle you plot. As long as the center points are perfectly aligned, you can project the plot line to any circumference (larger or smaller) (Fig. 128).

Actually the most practical scale would be one with a complete breakdown of 100 equal parts. Using such a scale would eliminate the repositioning to find the various points (Fig. 129).

You can, of course, prepare your own scale as just described, or you might simply save yourself the trouble and trace the scale on page 126. You can trace it either on vellum or on clear, prepared acetate (the latter would be much better). An even simpler way would be to have the page photocopied onto film, having a *film positive* made (black lines on clear film) to any size you choose. This would give you the most durable scale. However you prepare it, it would be worth the time and expense. There is really no need to prepare a scale every time you do a pie chart. This one scale will last a long time.

FIG. 128
continued

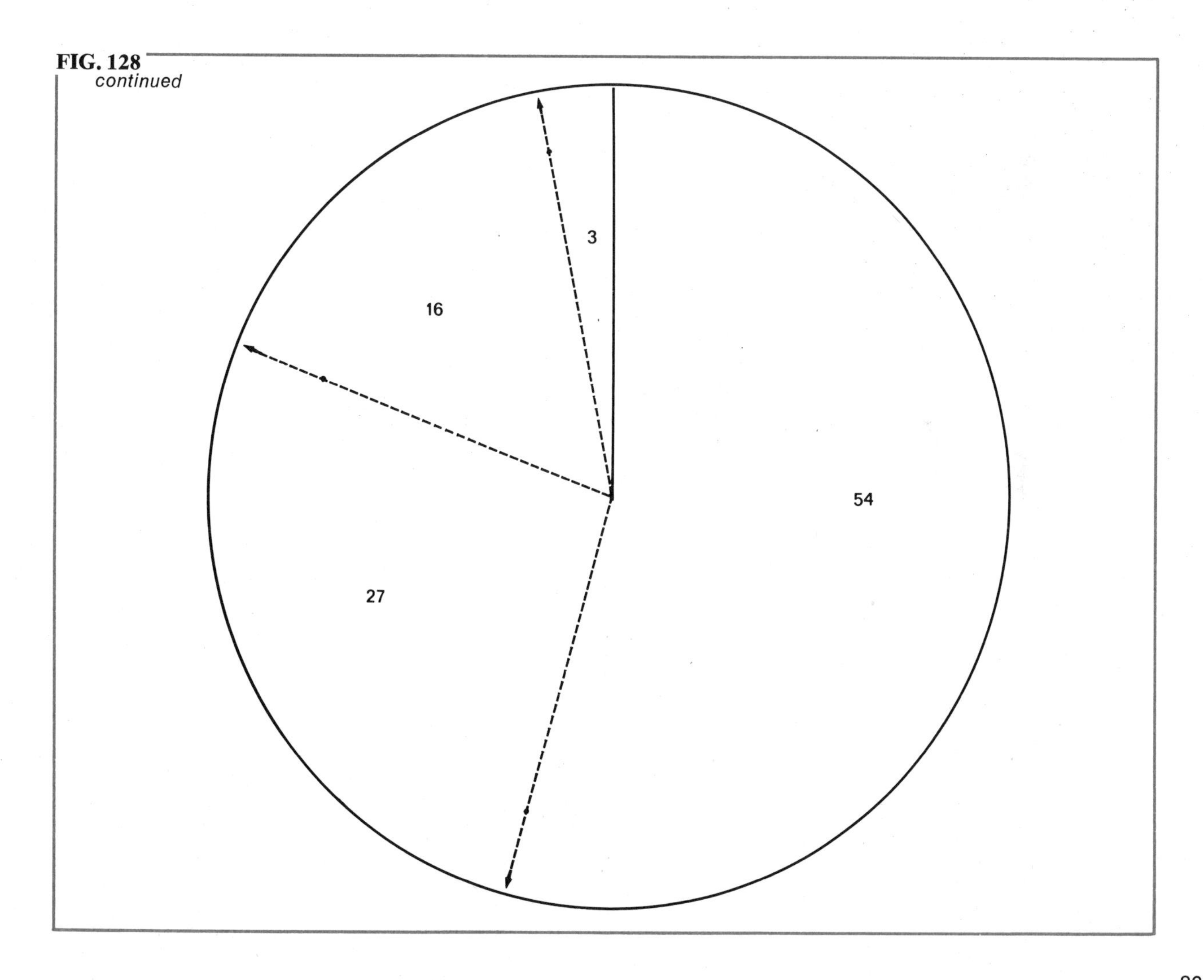

FIG. 129

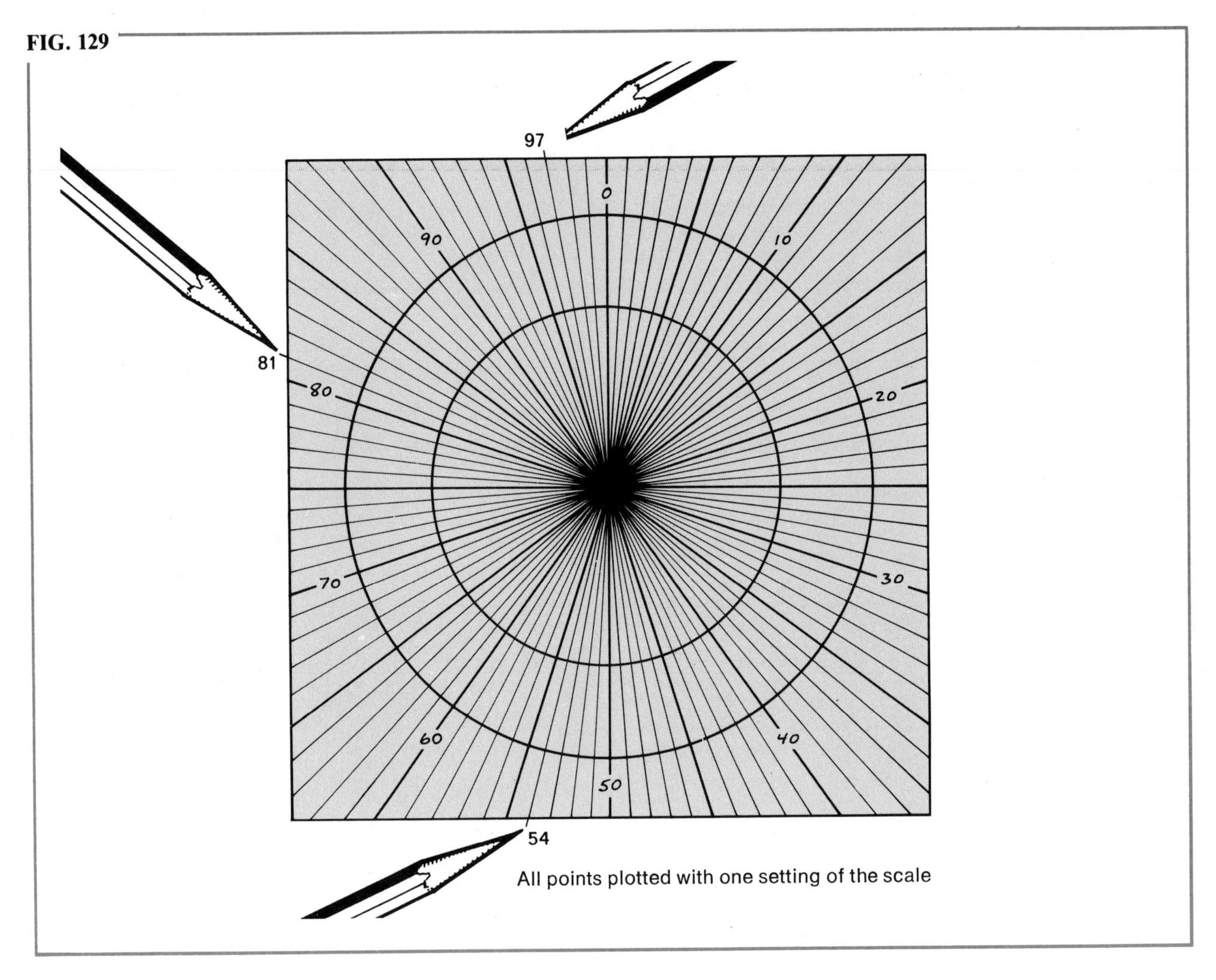

All points plotted with one setting of the scale

Exercise

Prepare a pie chart.
Diameter: $2\frac{1}{2}''$
Pie segments: 9%, 6%, 65%, 4%, 16%

Answer on page 125.

Chapter Six
A Direct Approach

QUICK PLOTTING

If you need to plot only two, three, or four very simple figures, it may not be necessary to prepare a grid scale.

Find the proper section on whichever ruler you have selected to use in preparing a particular grid scale. Then use the ruler itself directly in position on the chart as you would a grid scale.

For example, the chart in Fig. 130 is ready to be plotted. The figures are 85 for 1970, 42 for 1975, and 136 for 1980 (Fig. 130).

Use your engine divided ruler for this. Along the 20 edge you can use the first 10 increments (Fig. 131).

FIG. 130

FIG. 131

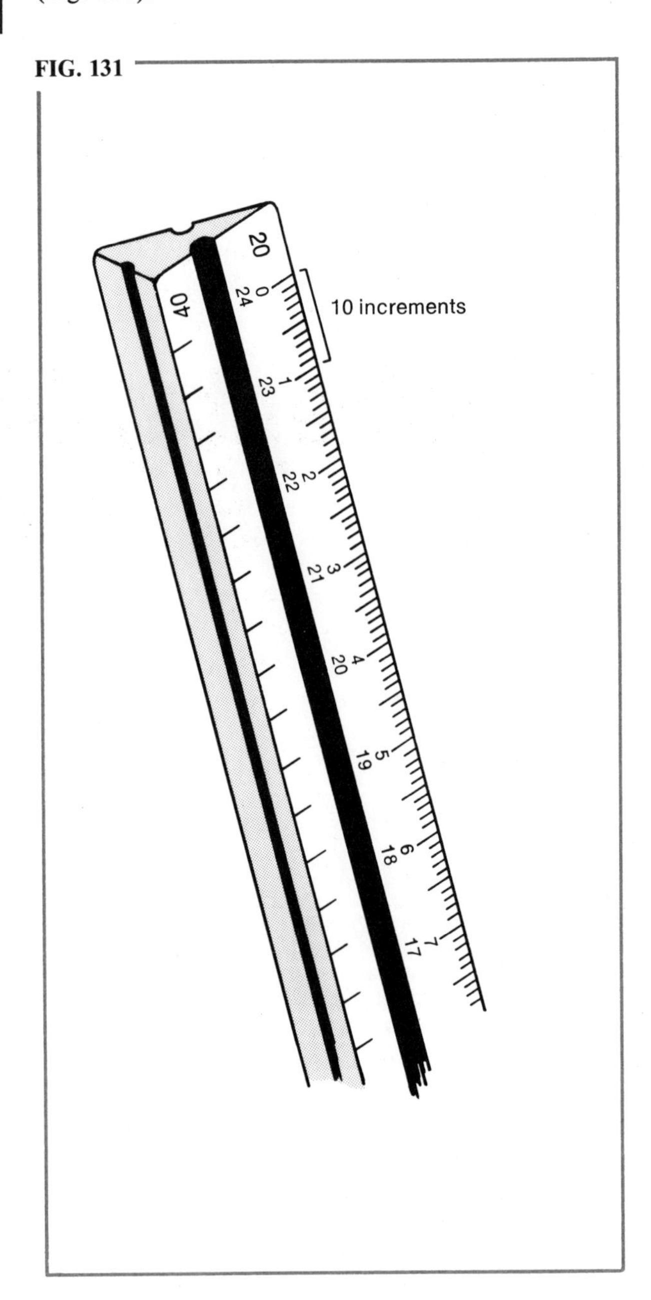

To plot the first figure (85), simply place the 0 mark of the ruler on the 80 grid line within the first column. Angle it until the 1 mark touches the 90 grid line. It does not matter if it touches beyond the width of the column. Hold your ruler firmly in position and mark off the fifth increment (plot point) (Fig. 132).

Remove the ruler, set up your T-square, and draw a line across the plot point to close off the first column (Fig. 133).

FIG. 132

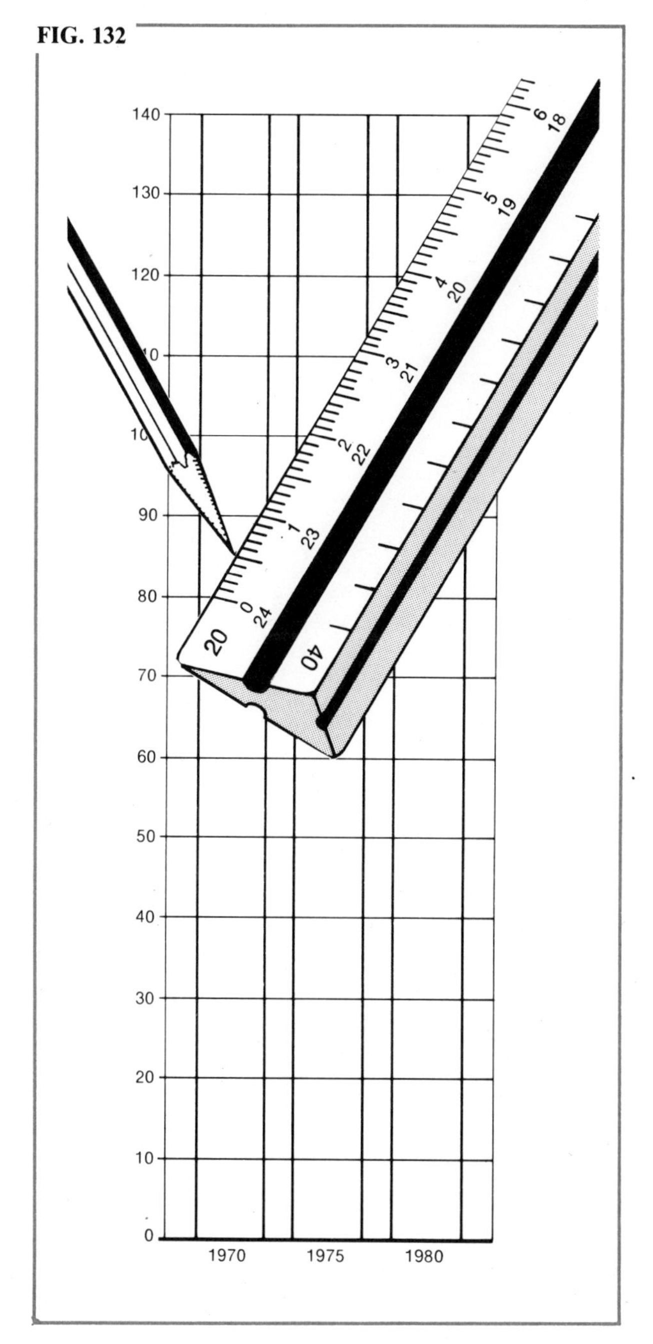

FIG. 133

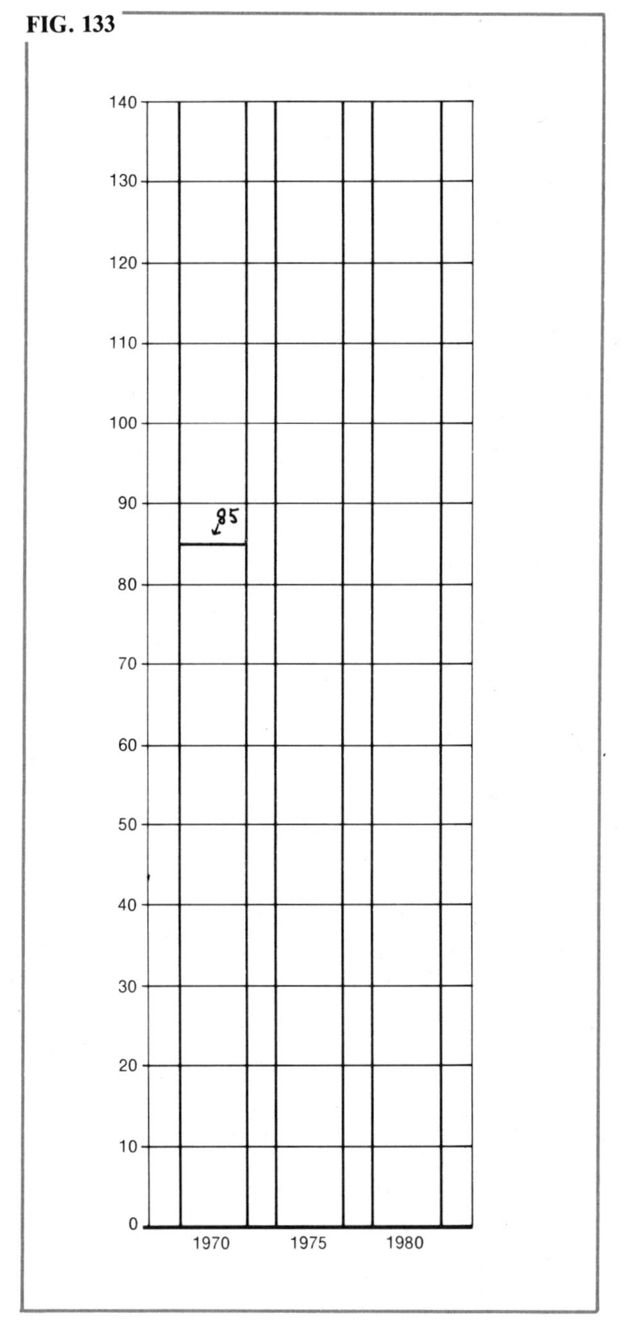

For the next figure (42), position *the same section of the ruler* on the second column between the 40 and 50 lines. Hold it firmly and mark off the second increment. Remove the ruler, set up your T-square, and draw a line across the column through that plot point (Fig. 134).

FIG. 134

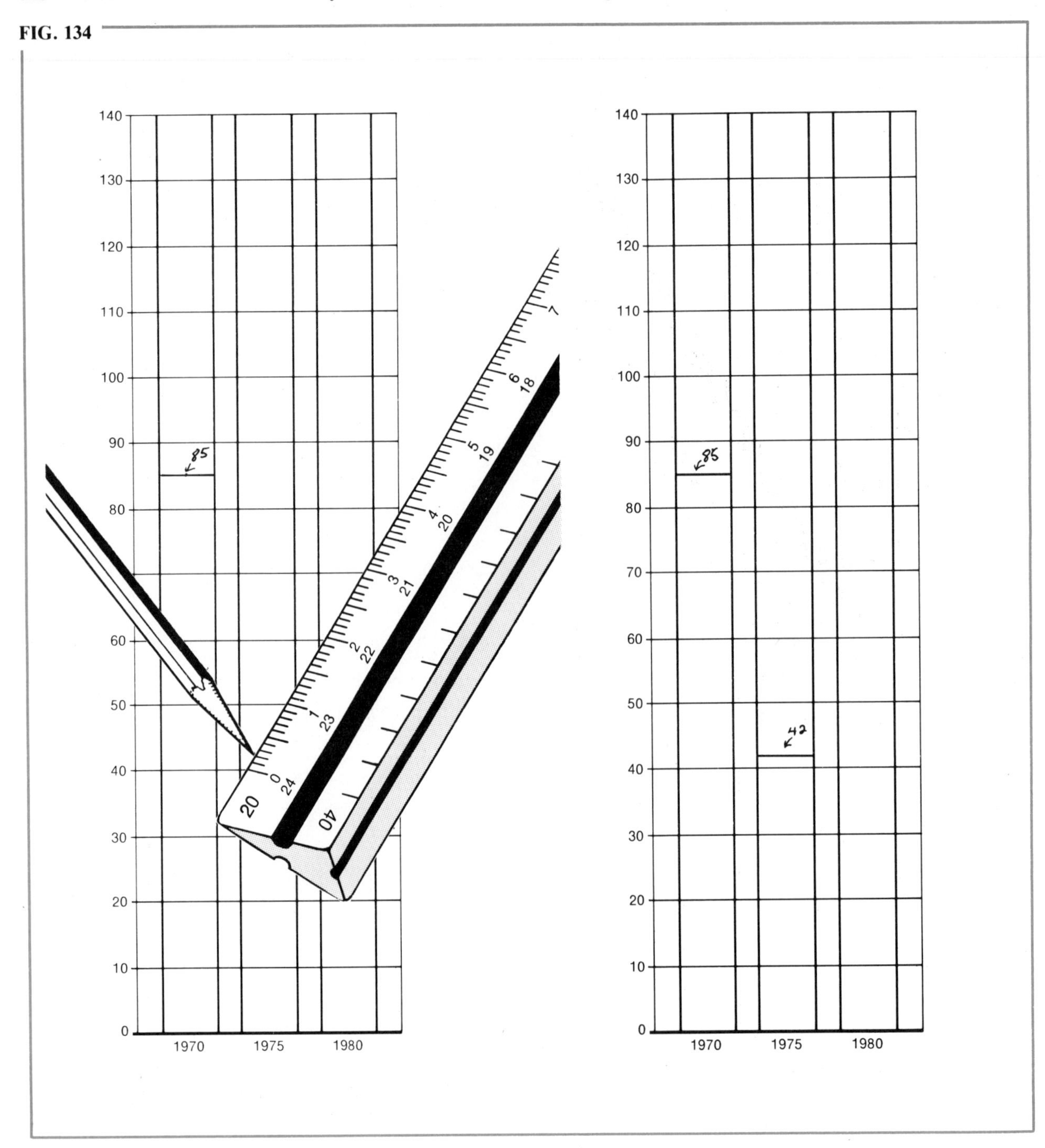

Repeat this procedure for the 1980 column (Fig. 135).

This approach can be used for line and layer charts as well.

FIG. 135

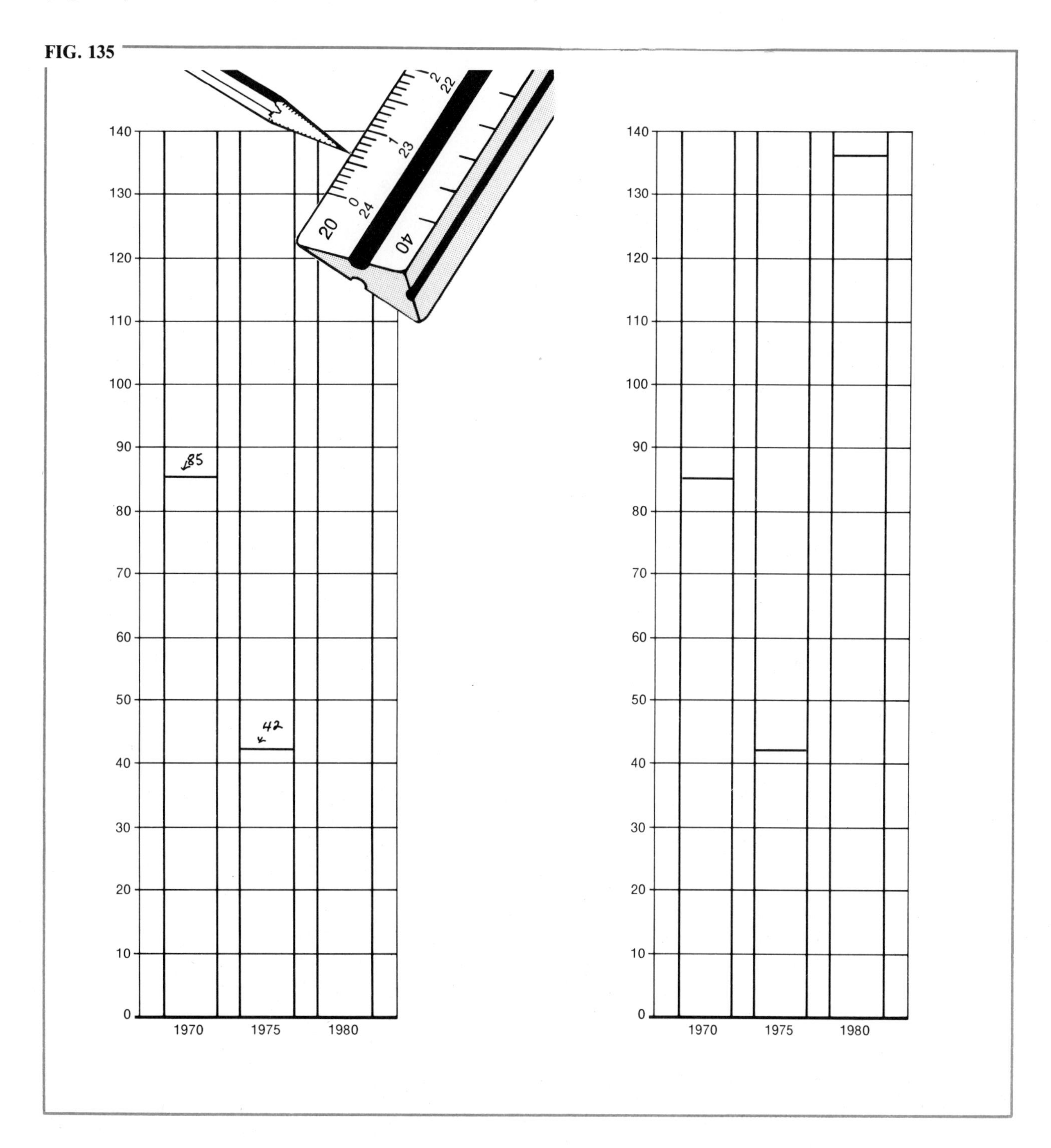

GRAPH PAPER

Most art supply stores carry pads or loose sheets of graph paper.

The graph pattern is an overall grid of uniform squares created by equally spaced horizontal and vertical lines. They are available in a variety of sizes and patterns and are printed in light blue ink. This makes it easier to see your drawing over the grid pattern (Fig. 136).

This paper is quite useful for planning most charts. You do not need to use a T-square: all you need are a pencil and a triangle to draw straight lines. The grid itself is actually the grid scale. By breaking down the number of squares to a specific ratio you can easily lay out most charts.

An $8\frac{1}{2}'' \times 11''$ sheet of the graph patterns shown in Fig. 136 is composed of 68 horizontal boxes and 87 vertical boxes. These boxes can be used to represent almost anything you choose.

Rather than working with actual mathematical dimensions, you work with numbers of boxes (or *lines* if you wish). Let's see how the graph pattern is used.

The specifications are as follows:

Production Comparison Between Company A and Company B Over a Period of Ten Years
In Hundreds

	A	B		A	B
1971	148	100	1976	167	145
1972	127	158	1977	155	210
1973	134	158	1978	194	215
1974	152	152	1979	210	195
1975	160	150	1980	235	240

FIG. 136

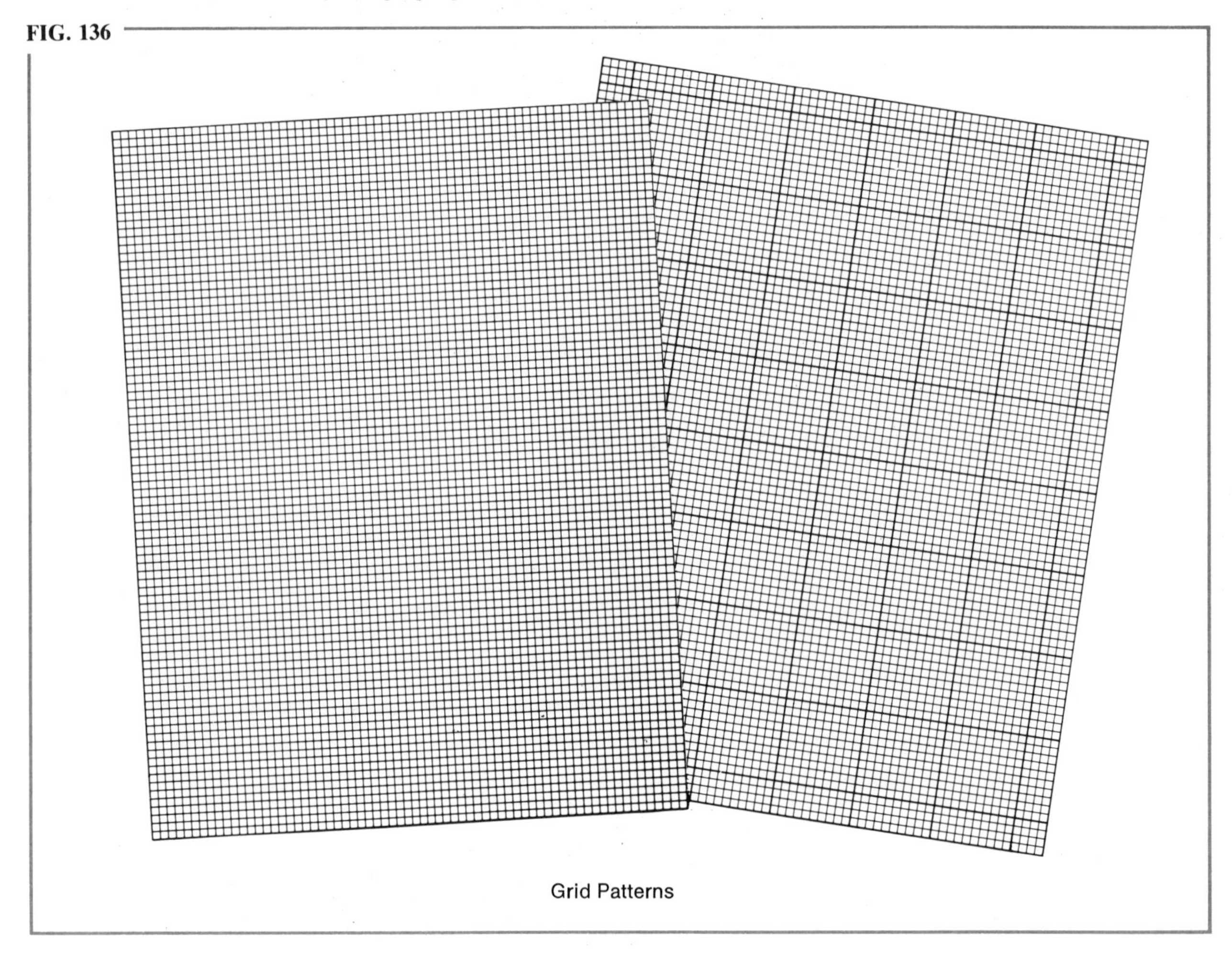

Grid Patterns

The chart itself will be a line chart requiring 9 divisions on the width, representing the years 1971 through 1980. Think about this for a moment. There are 10 dates. Thus you will need *10* vertical *lines* but only *9 spaces*. The first line is 1971, and the last is 1980. Count the *spaces* from the first to the last line; there are 9 spaces.

Before you start your sketch, take the time to determine the number of boxes you will need for the width and height. You might use one box for the space between successive years, but that may be too small to be practical. Why not 3 or even 4

FIG. 137

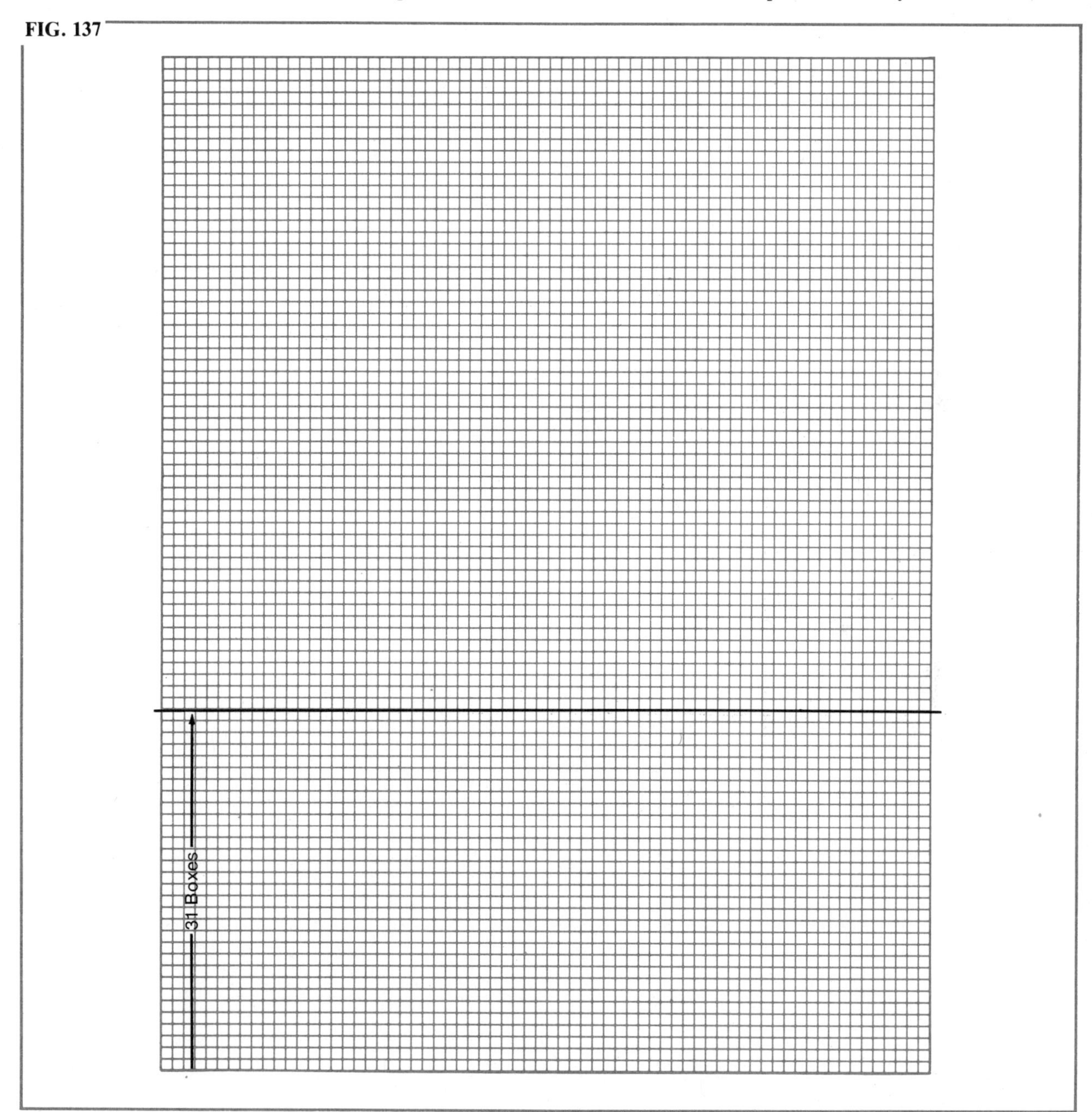

boxes between year lines? This would give 4 × 9 = 36 boxes on the width.

The highest figure to be plotted is 240. The chart could end at 250. Your scale in multiples of 50 would require a total of 5 increments. Each increment will be 5 boxes high, giving a total of 25 boxes on the height.

Thus your overall dimensions are 36 boxes wide by 25 boxes high. Pencil in those dimensions on the graph paper. When preparing this sketch you might take the time to center it on the page. To do so, simply subtract the height (25 boxes)

FIG. 138

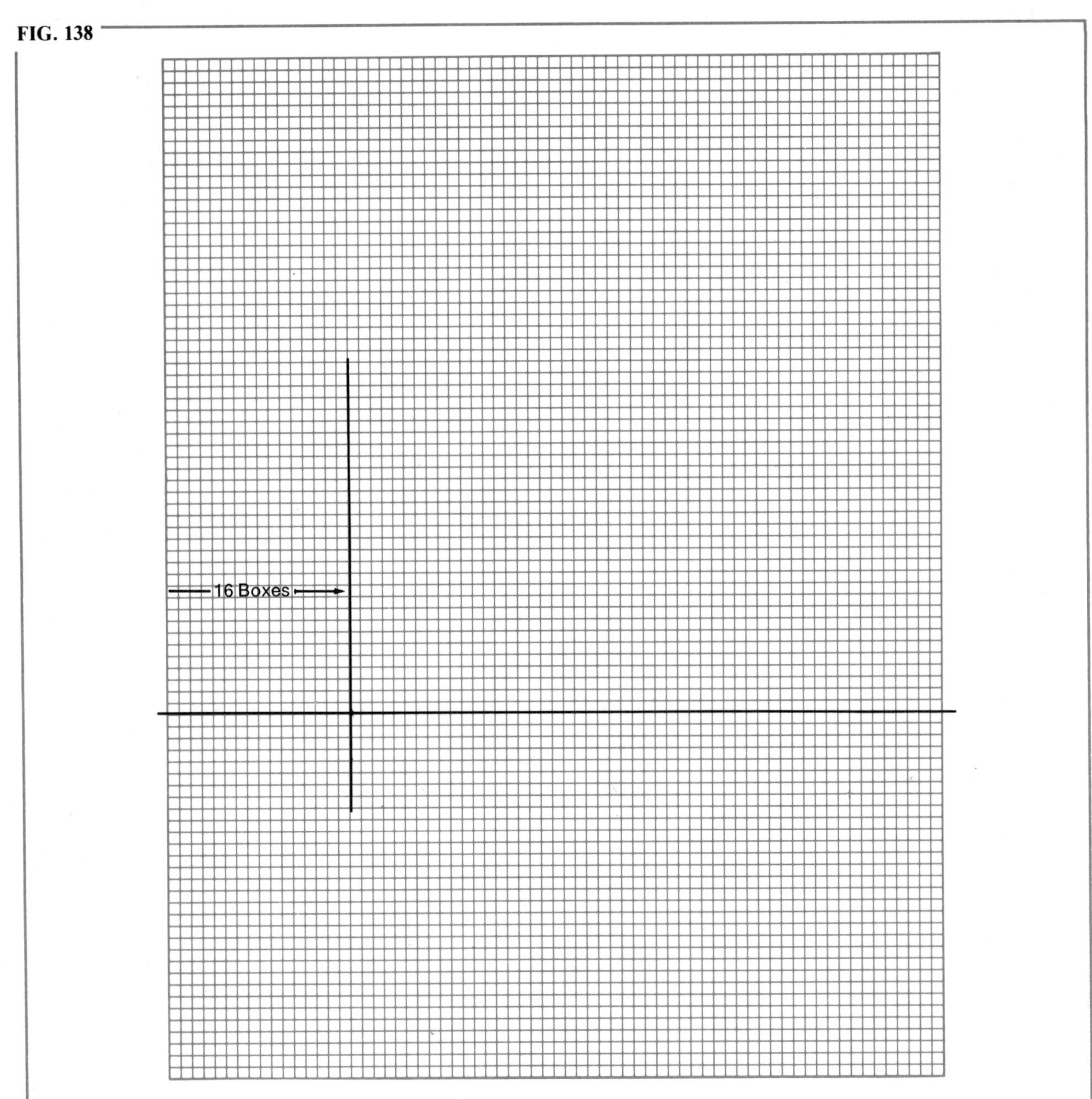

from the total number of boxes in one row on the height of the graph (87 boxes), which leaves 62. Divide 62 by 2, obtaining 31. Count up 31 boxes from the bottom; that is where you will draw your base line for the chart (Fig. 137).

Do the same for the width. You need 36 boxes for the width. Subtract 36 from the number of boxes in one row on the width of the graph: 68 − 36 = 32. Divide 32 by 2: 32 ÷ 2 = 16. Count 16 boxes from the left; that is the beginning of the base line (Fig. 138).

Now count off 36 boxes along the base line

FIG. 139

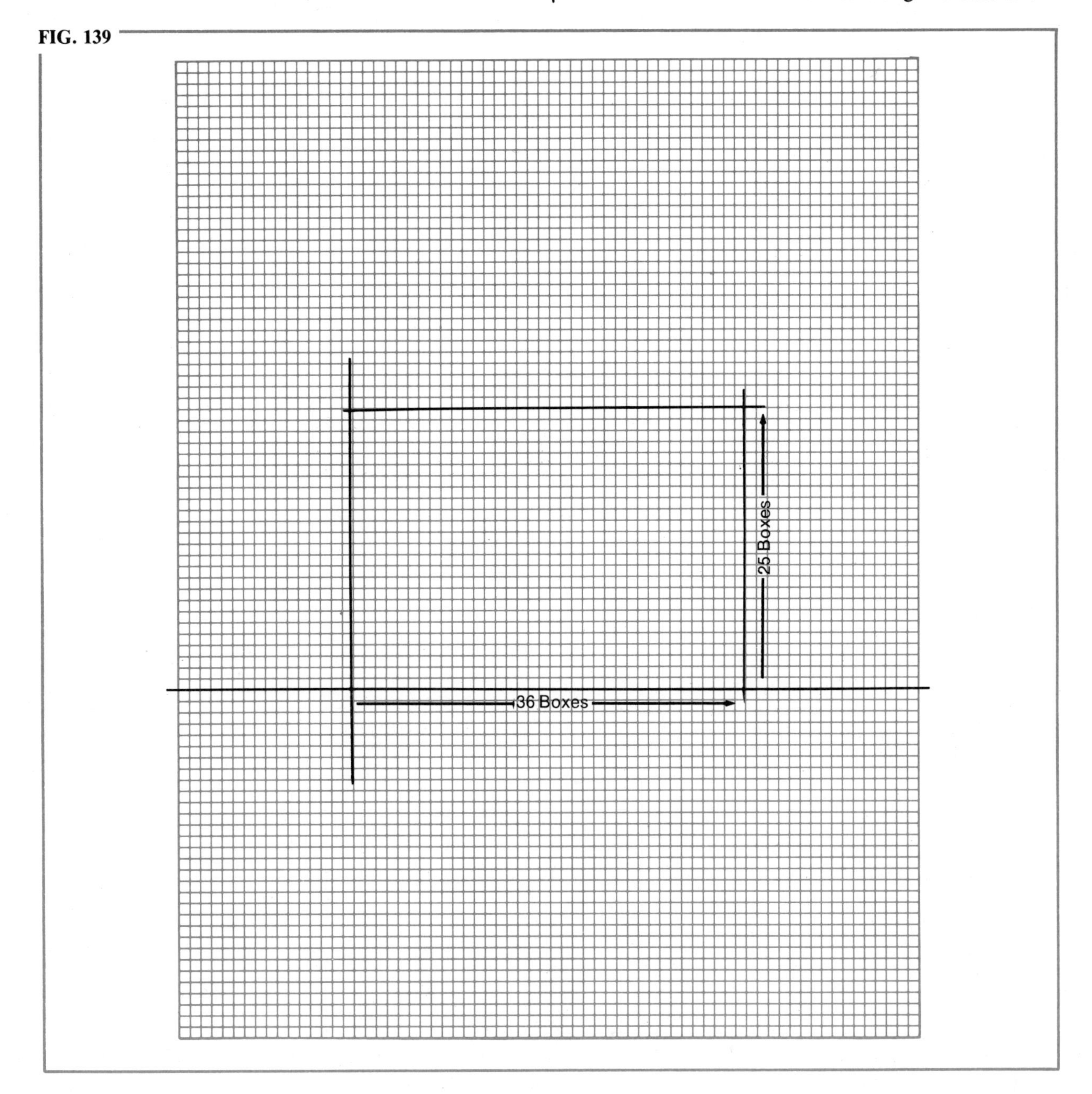

to define the overall width. Then 25 boxes up to establish the overall height (Fig. 139).

Pencil a line horizontally across at every fifth box to establish the grid and vertically at every fourth box to establish the year lines. Label all the lines (Fig. 140).

You are ready to plot the figures, but you do not need to prepare a grid scale. It is already superimposed over the entire chart. Look at the first vertical line for 1971. Each 50-unit increment is composed of 5 boxes. Each box represents 10. The first figure in Company A is 148. Count the

FIG. 140

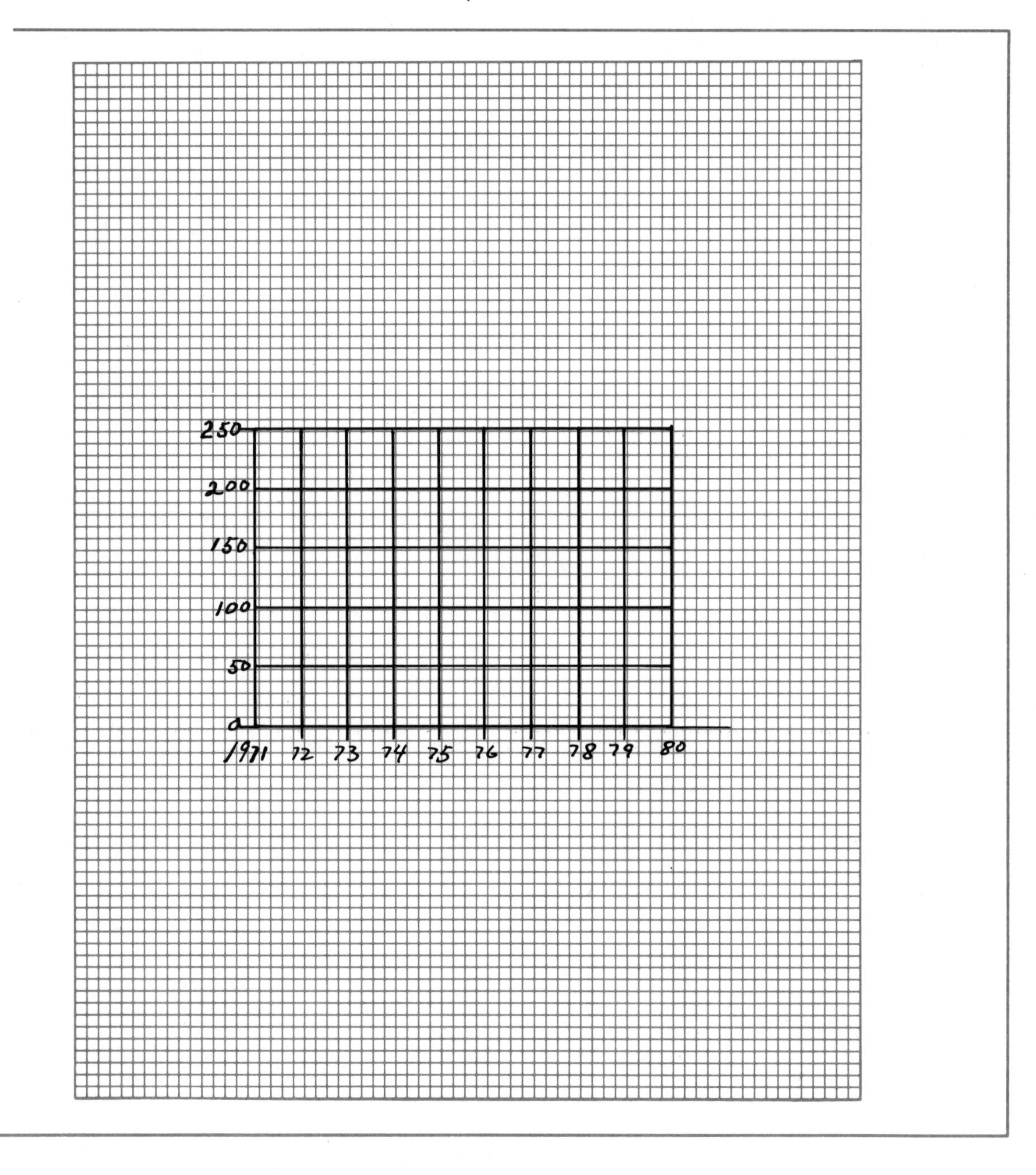

boxes up from the 100 grid line to 140, and mark off by eye a little above the center of the fifth box for 148 (Fig. 141).

Repeat this with each figure along each year line. When you connect the plot points, your layout should look like Fig. 142.

Try it yourself, using the same specifications in a horizontal bar chart. The thickness of each bar is 2 boxes, and the space between groups is 1 box. The distance between the year lines is 10 boxes, each representing 5 units. The result should look like Fig. 143.

FIG. 141

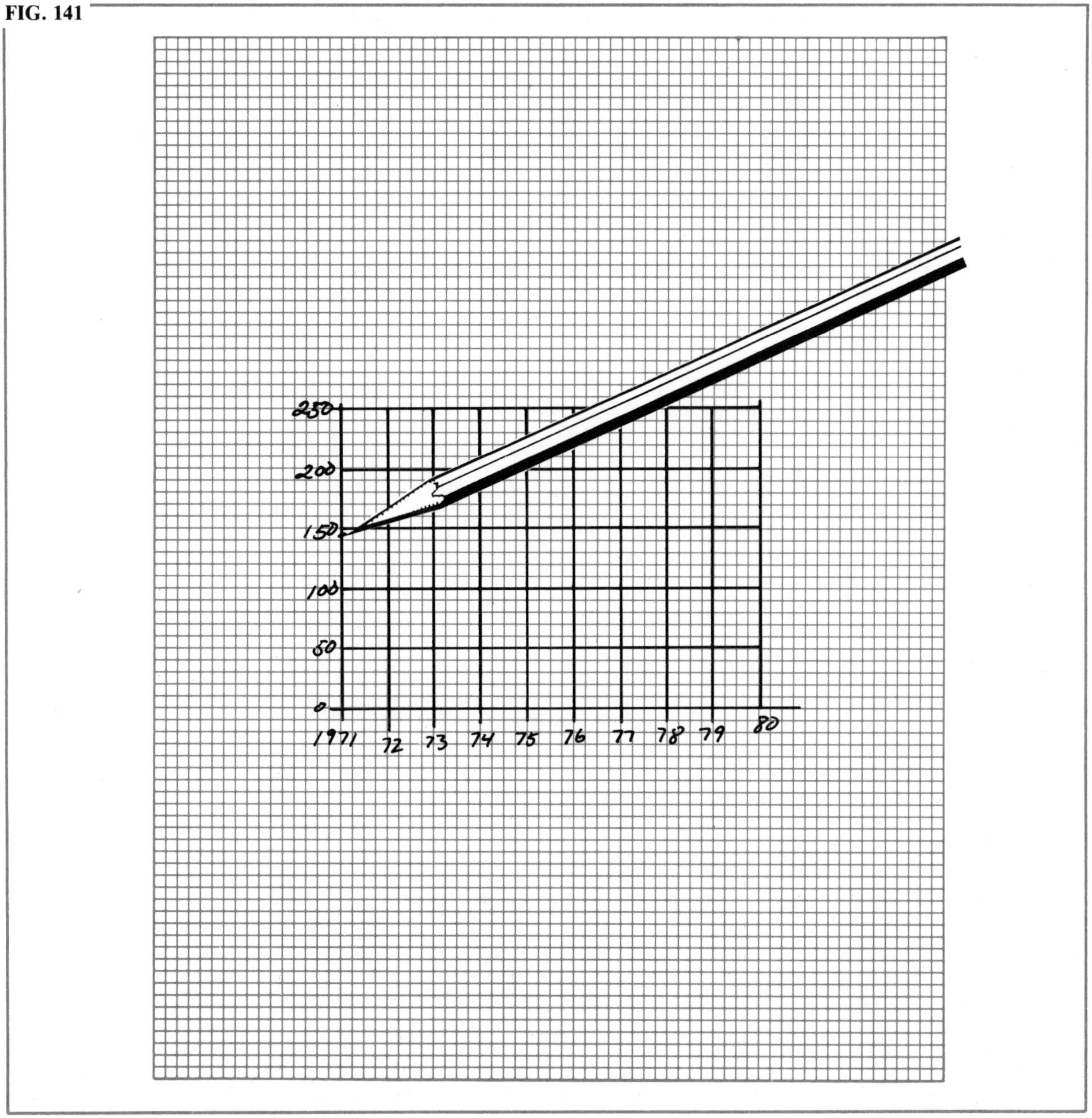

FIG. 142

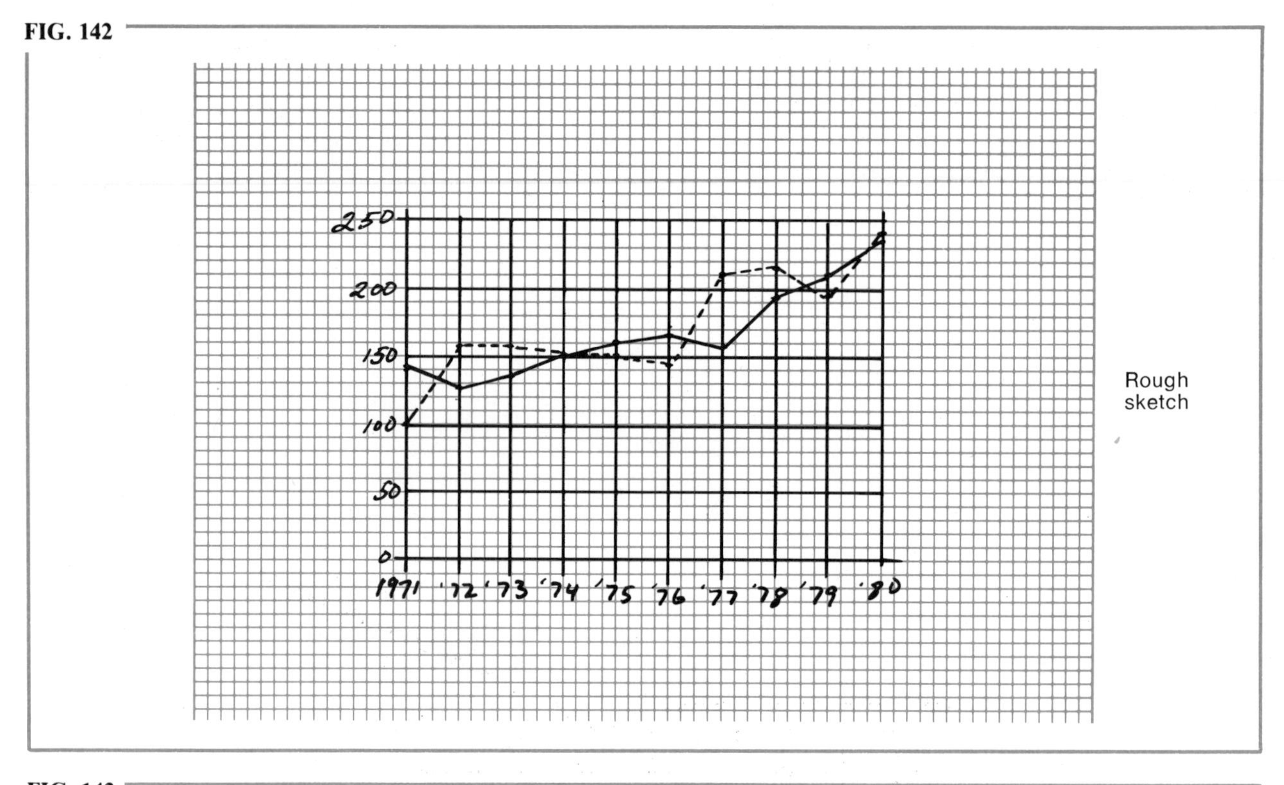

Rough sketch

FIG. 143

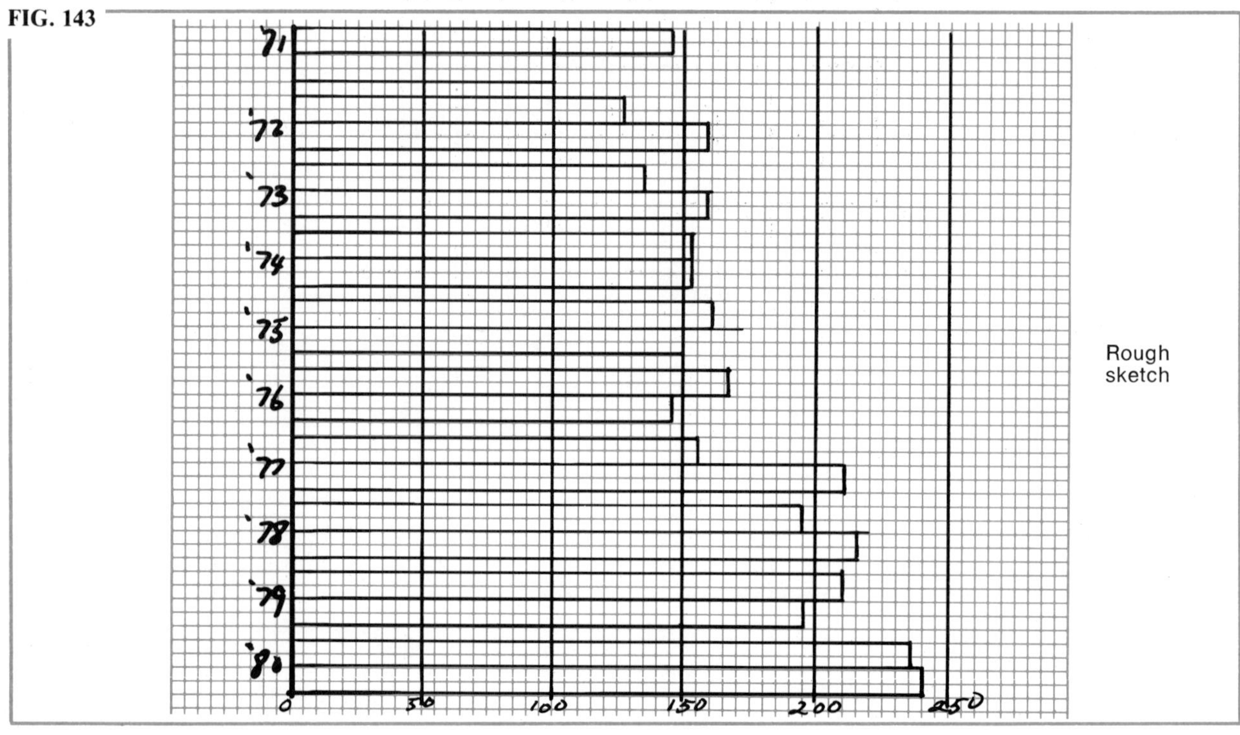

Rough sketch

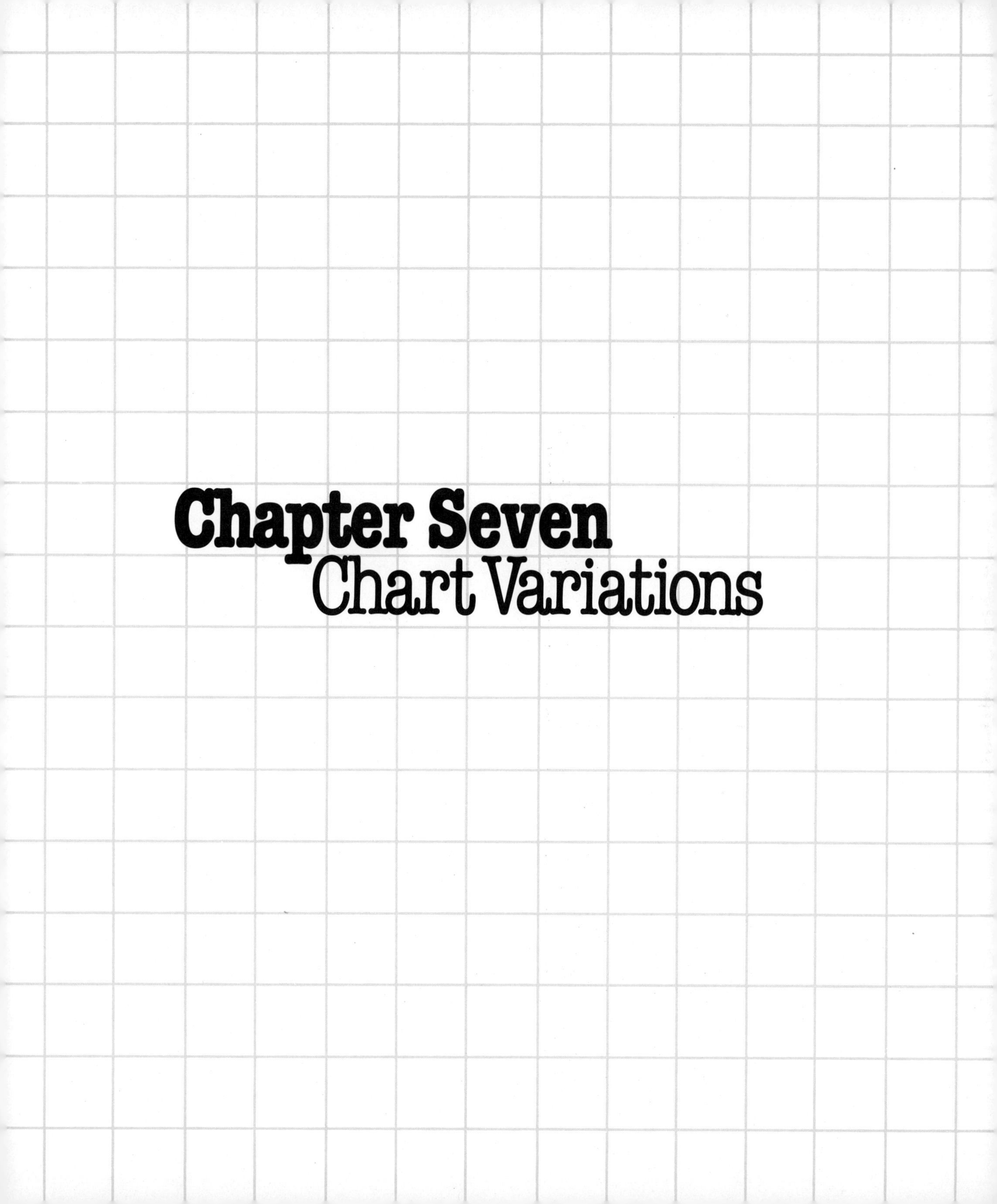

Chapter Seven
Chart Variations

THE HIGHEST POINT

In the examples and exercises throughout the text, the top line, or the highest point of the chart, was rounded off from the highest plot figure. But suppose the requirements were such that the highest point of the chart had to be the highest *plot* point. How would you determine that point and the grid divisions for the scale?

It would be simple enough if you used graph paper. After determining the scale by letting a certain number of boxes be equal to a specific scale, you would mark off the highest figure, and that would be the actual end of the chart. For example, work with a vertical column chart that has $1,854 as the highest plot point. Under normal circumstances, you would round off $1,854 to $2,000 and prepare a scale with 8 increments of $250 each. But in this instance you will use only the first 7 increments, each 5 boxes high. There would then be a total of 35 boxes to the $1,750 line, each box representing $50. To find $1,854, count 2 boxes up from the $1,750 line to $1,850 and mark off, by eye, 4—just a bit over the line. That point would then be the absolute height of the chart (7 increments, plus the highest plot point) (Fig. 144).

FIG. 144

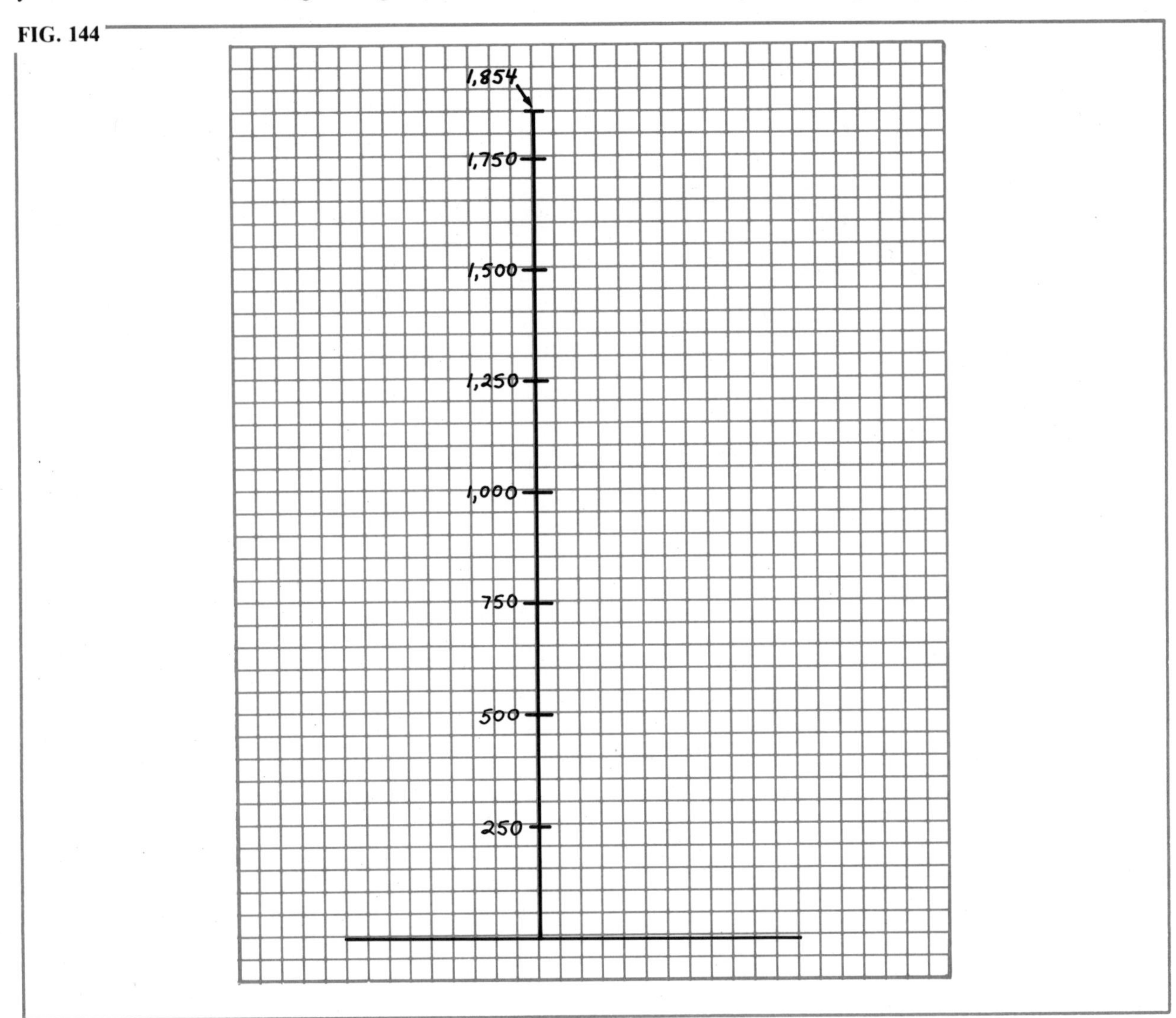

This is relatively simple if you are working with graph paper and have the freedom to prepare the chart to an arbitrary size. But suppose you were restricted to a specific height, say, $3\frac{7}{16}''$? You don't have the convenience of the boxes, but you do have all you have learned in this book so far.

The specified height of $3\frac{7}{16}''$ is actually the $1,854 plot point. In a way, this is working backwards. You start out with a predetermined highest point and must establish the scale between it and the base line.

It really isn't difficult to work out. Construct the dimensions of the chart ($3''$ by $3\frac{7}{16}''$). Then determine which ruler you want to use. Let's use the engine divided ruler. We have already decided on 7 increments plus the plot point. Look at the 10 edge of the ruler. The 4″ mark could be used. Counting every fifth increment ($\frac{1}{2}''$ mark) would give you 8 spaces, each broken down into 5 parts. Now study your ruler for a moment. Find 1,750 (the 7th $\frac{1}{2}''$ mark). From the 7th mark count the small subdivisions (50 each): 1,800, 1,850 and just a hair beyond that point should be 1,854 (Fig.145).

Position the 0 mark on the base of the chart, and pivot the ruler until the 1,854 mark touches the top line (Fig. 146).

FIG. 145

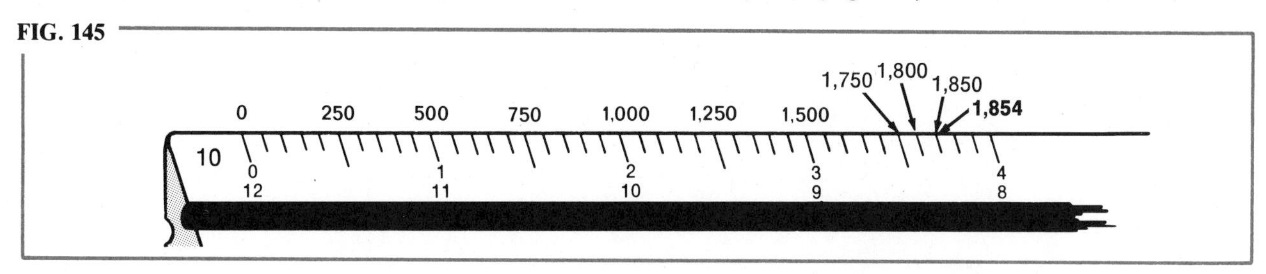

FIG. 146

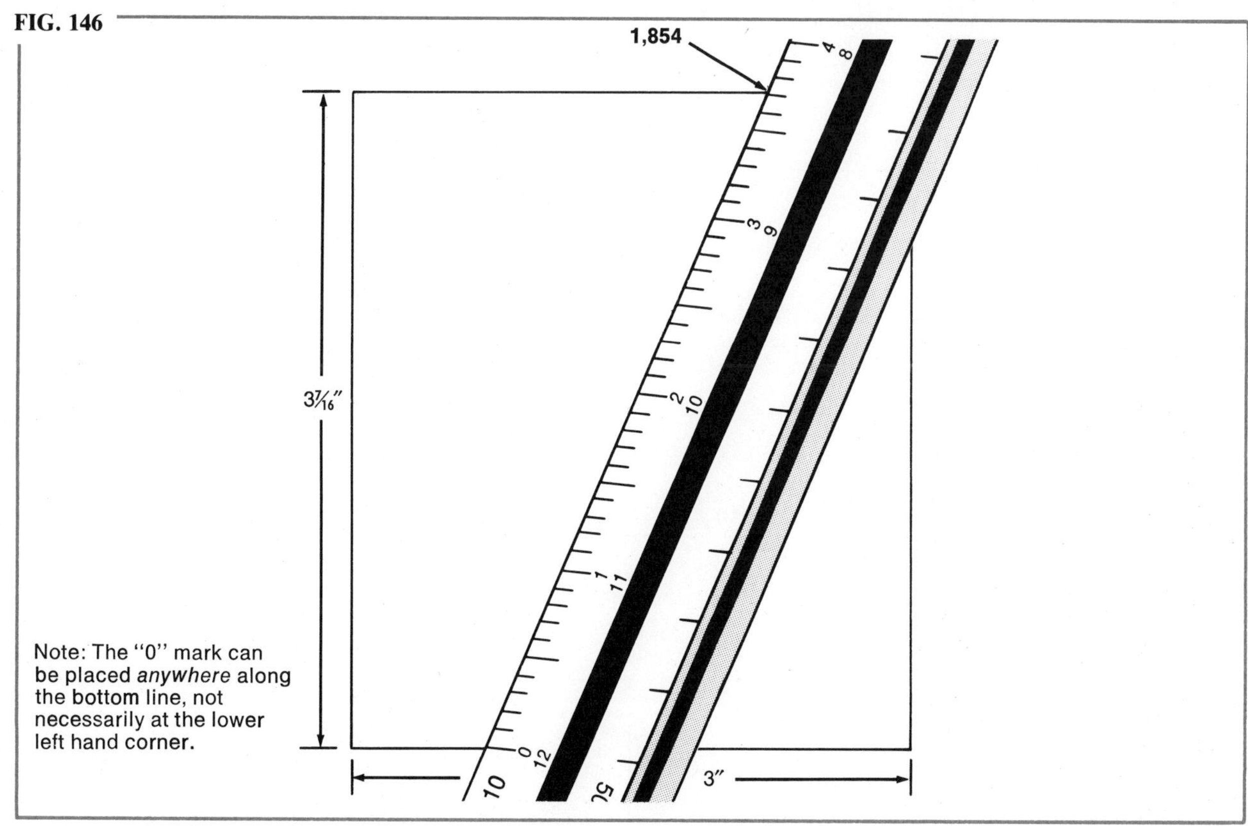

Hold your ruler firmly in this position and tick off every $\frac{1}{2}''$ mark (for a total of 7). Rule lines across at each tick mark; you have created a grid with 1,854 as the highest point within a specific area (Fig. 147).

FIG. 147

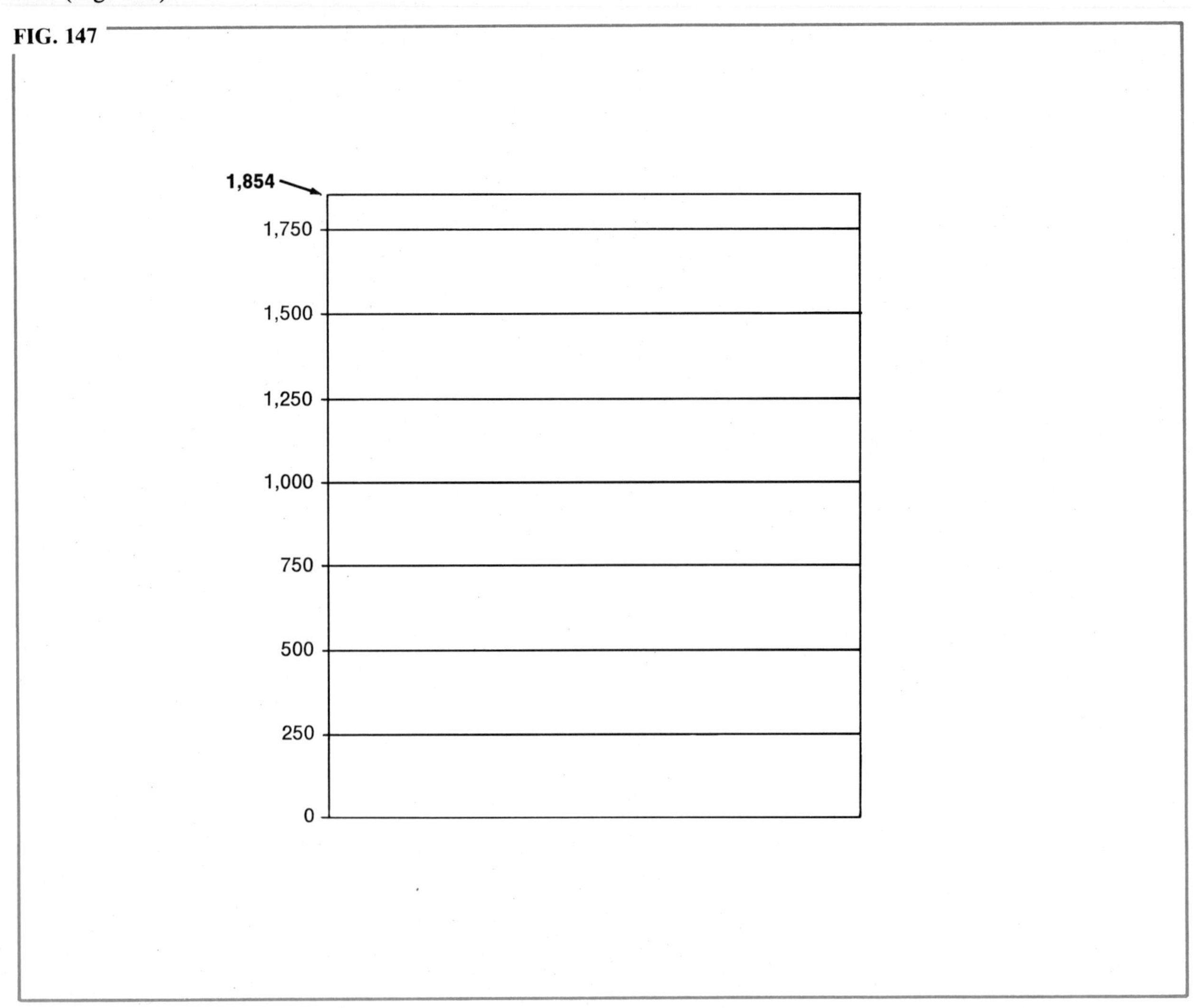

Exercise A

Type of chart: column
Overall dimensions: $2\frac{7}{16}''$ by $3\frac{1}{8}''$
Highest plot point: 913 at $3\frac{1}{8}''$
Number of grid increments:
9 + the 913 plot point
Each grid line: 100

Answer on page 125.

Exercise B

Type of chart: horizontal bar
Overall dimensions: $1\frac{7}{8}''$ by $3\frac{1}{8}''$
Widest plot point: 536 at $1\frac{7}{8}''$
Number of grid increments:
10 + the 536 plot point
Each grid line: 50

Answer on page 125.

SUBDIVIDED COLUMN CHART

In Chapter 4 you learned the principles of constructing a layer chart by cumulation of the figures. The same approach can be employed in the preparation of a bar or column chart.

The chart in Fig. 148 is ready to be plotted.

The component figures are:

	Major	R&D	OM	Construction
1977	24	8	12	5
1978	28	6	16	2
1979	21	6	11	3
1980	18	3	10	4

FIG. 148

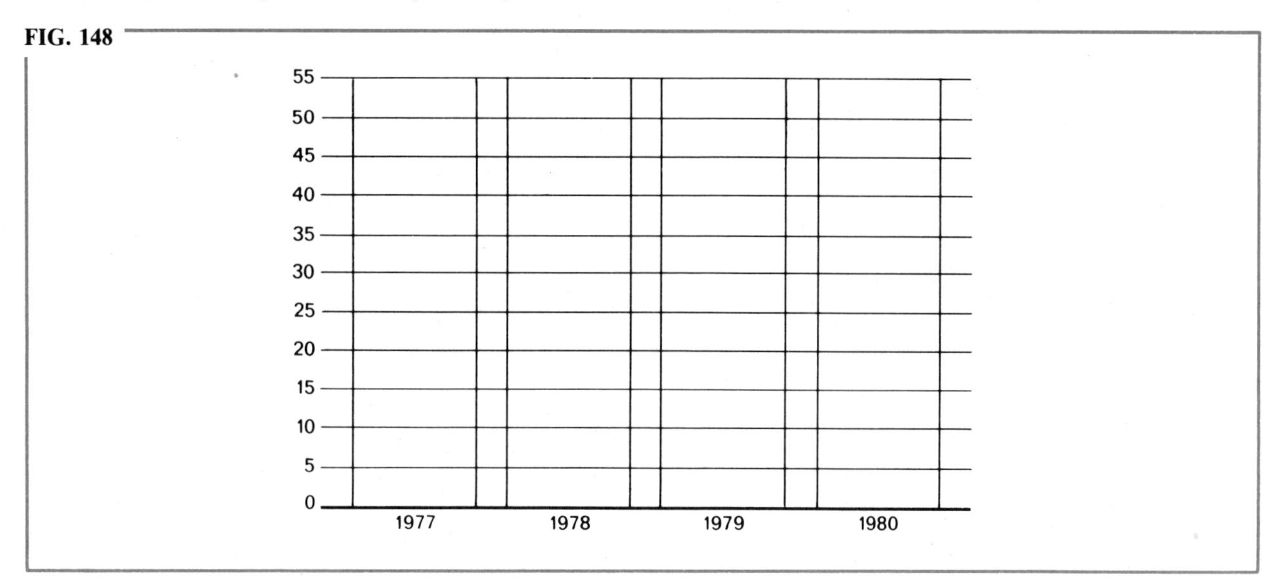

We'll start with Major Systems as the first component from the bottom up. First prepare a grid scale, plot *24* in the 1977 column, and draw a line across at that point (Fig. 149).

FIG. 149

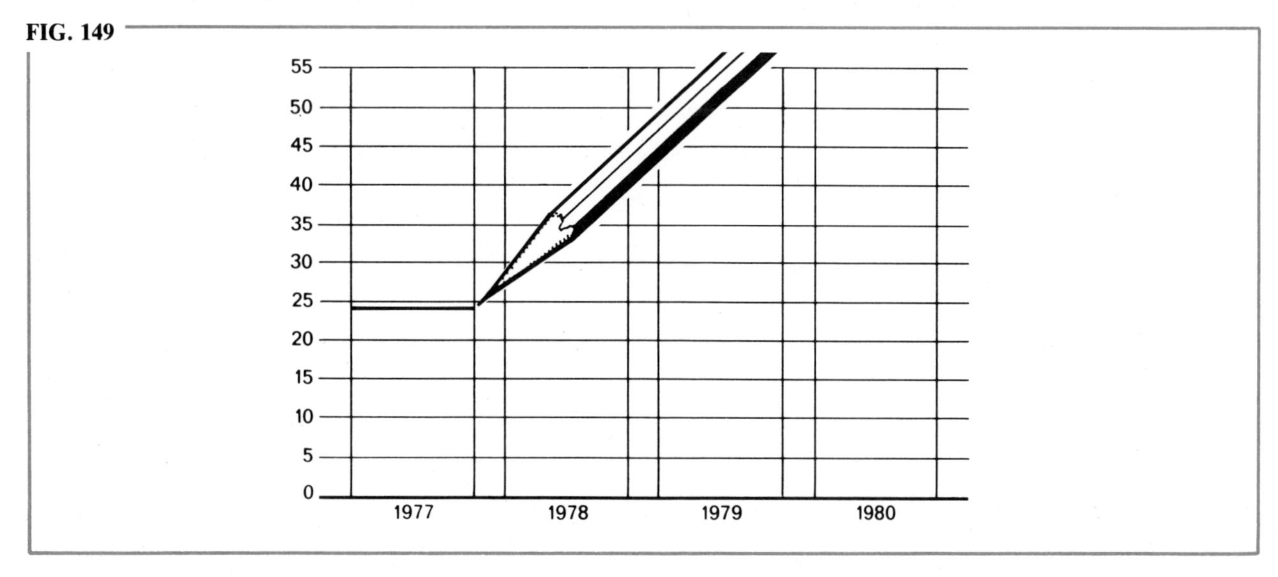

Now add the R&D figure (8) to the 24; 8 + 24 = 32. Plot *32* in the 1977 column and draw a line across (Fig. 150).

FIG. 150

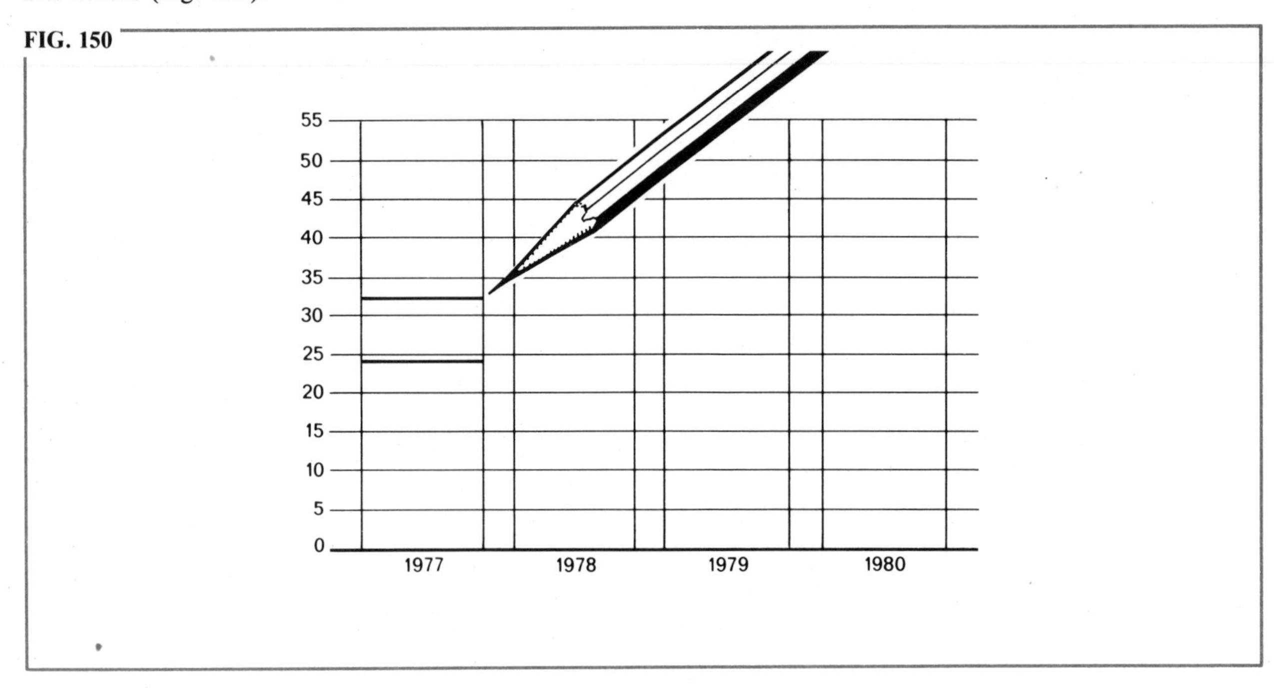

Add the OM figures (12) to the 32; 12 + 32 = 44. Plot *44* in the 1977 column (Fig. 151).

FIG. 151

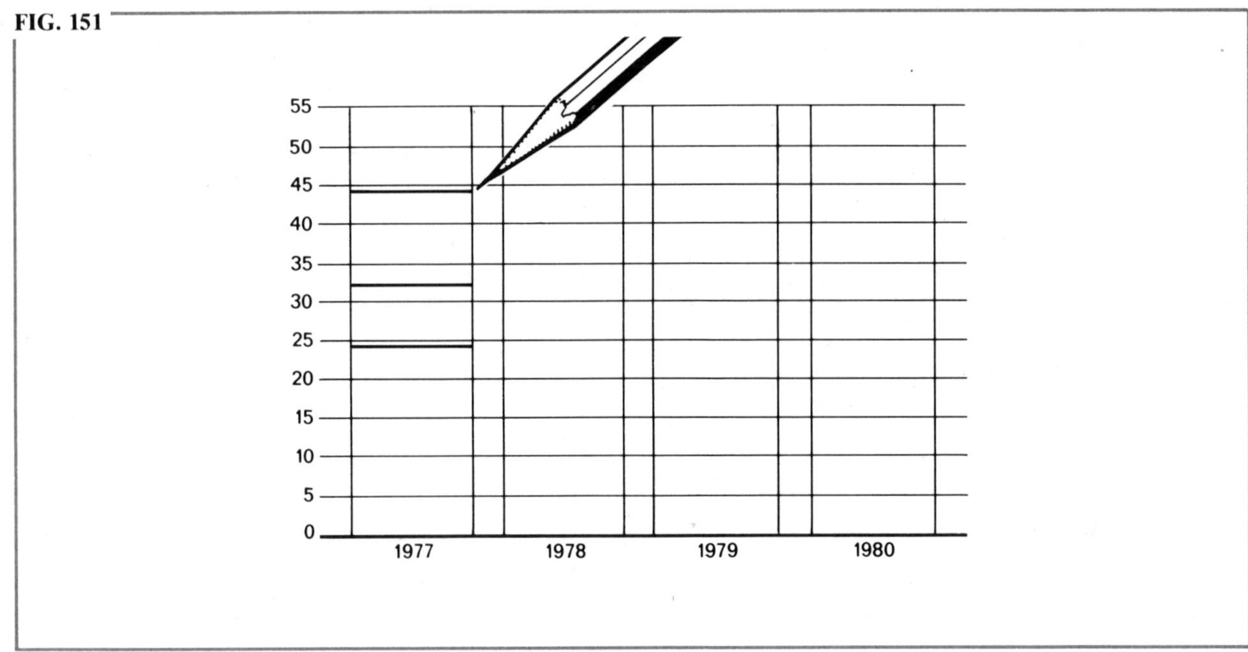

Add the Construction figure (5) to the 44; 5 + 44 = 49. Plot *49* and draw a line across to close off the top of the 1977 column (Fig. 152).

FIG. 152

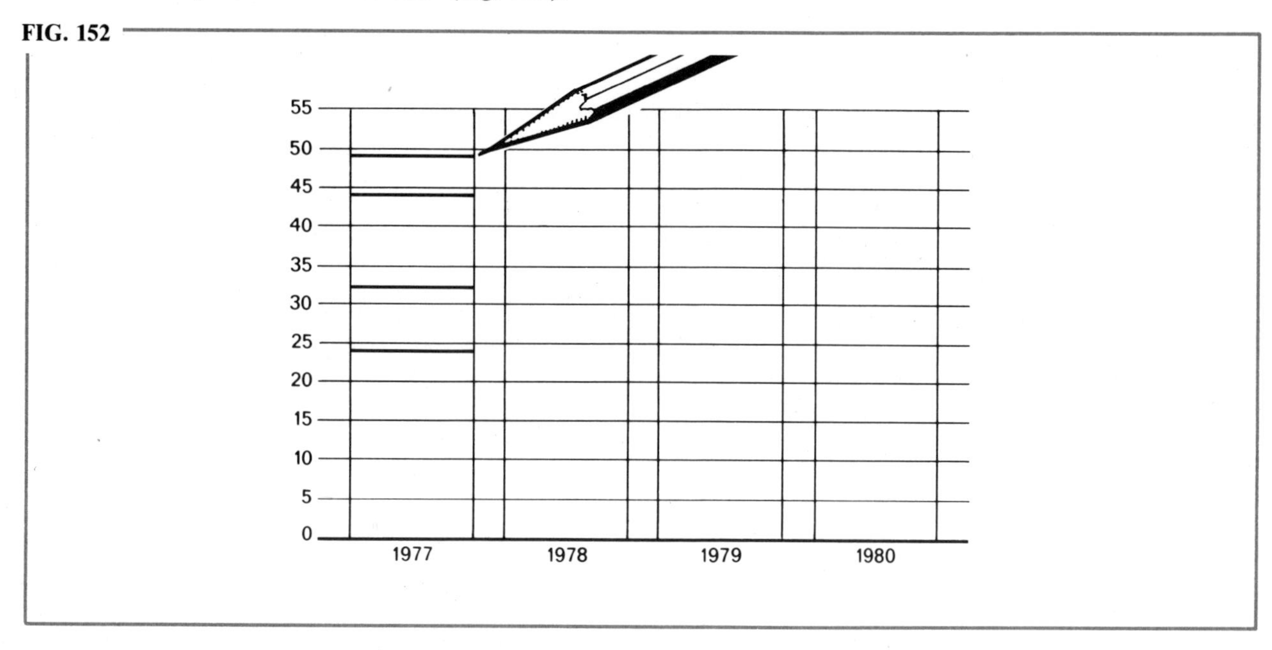

Follow the same procedure for the 1978 column. Plot the first figure (28). Draw a line across and add 6 to the 28; 6 + 28 = 34. Plot *34*, draw a line across, and add 16 to the 34; 16 + 34 = 50. Plot *50,* draw a line across, and add 2; 50 + 2 = 52. Plot *52*, ending the 1978 column.

Plot the remaining columns on your own. The result should look like Fig. 153.

FIG. 153

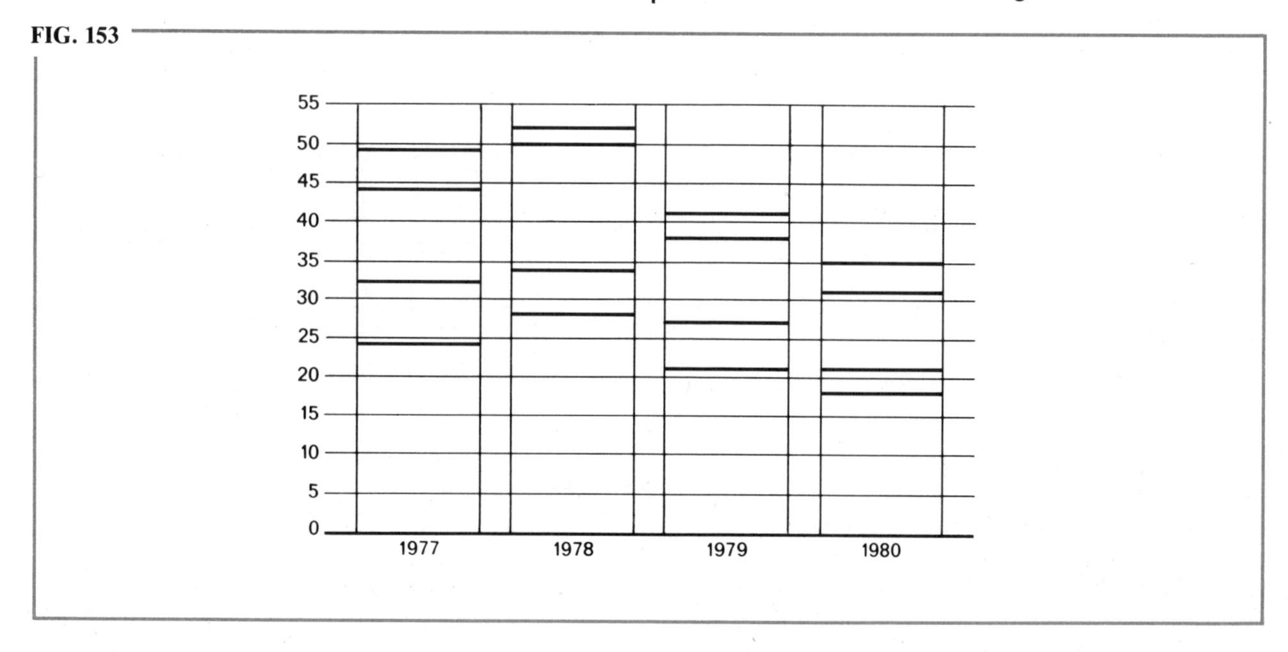

The use of different textures helps identify each component (Fig. 154).

FIG. 154

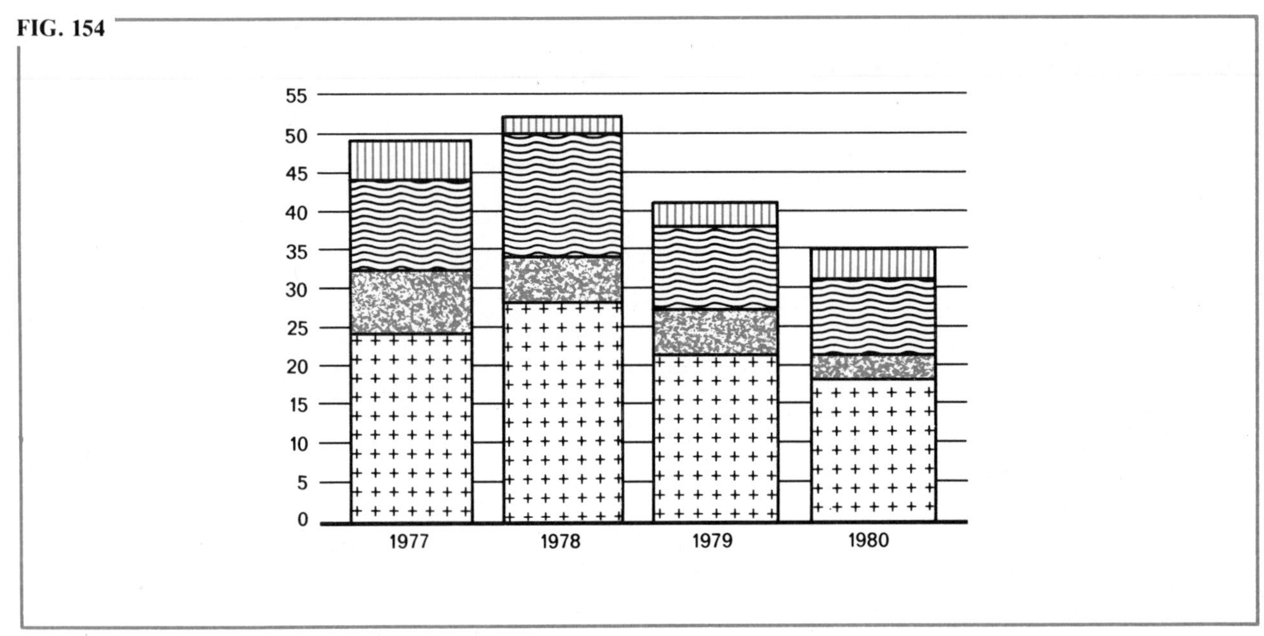

Test yourself with this one.

Exercise

Type of chart: 100% subdivided horizontal bar. Each bar is to have 5 components. The figures for each component are to be converted to percentages and plotted as layers (subdivisions) for each bar. The specifications are:

	Billions of Kilowatt-hours				
	Coal	Gas	Oil	Water	Nuclear
1965	50	30	10	22	0
1970	76	32	8	18	4
1975	92	32	8	34	38
1980	95	32	9	41	95

Overall dimensions: 18 picas wide by $10\frac{1}{2}$ picas deep. Thickness of each bar: 2 picas.

Answer on page 125.

THE SPLIT GRID

On occasion you will be presented with the problem of illustrating figures having a relatively narrow range but falling very high on the scale, as in Fig. 155.

FIG. 155

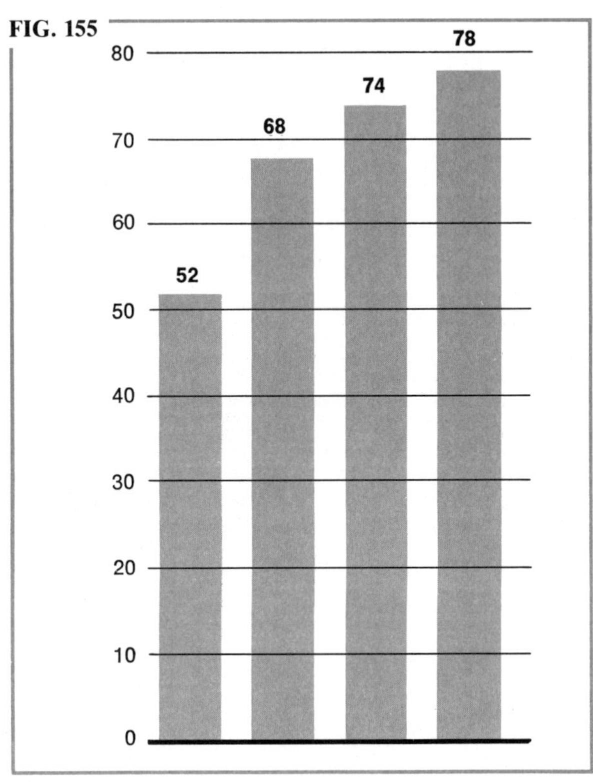

Note that the lowest figure is quite high on the scale and the highest figure isn't very much higher. The relationship of one bar to the next doesn't seem particularly dramatic. Actually, all that space at the bottom of the chart is wasted.

One solution is to split the bottom section in such a way that a portion of it can be removed. The top part of the chart could then be moved down and expanded upward again to the original height. This would result in wider increments.

While the heights of all increments would be equal, the first one would represent more units than the others. This is suggested by the "split" or separation of the columns within that first increment in Fig. 156.

The numbers plotted in Fig. 155 and Fig. 156 are the same, but because there are fewer and deeper increments, they contrast more dramatically in Fig. 156.

Another approach is to eliminate the base 0 line altogether and let the base line be the lowest increment figures below the lowest plot figure, as in Fig. 157.

It should be apparent that there are no figures lower than 50. This style focuses more attention on the relationship of the figures. The space between increments in Fig. 156 and Fig. 157 is also much wider than in Fig. 155, offering better readability and easier, more accurate plotting.

NEGATIVE FIGURES (DEVIATION-COLUMN CHART)

Negative figures are figures below 0, and that is precisely where they will appear (Fig. 158).

The scale below the 0 base line is usually the same as above and is plotted in the same manner. A horizontal chart with negative figures would look like Fig. 159.

A variation on this type of chart would be to split a column into two components, one of which is above the 0 line and the other below. For example, the total length of a bar might represent a total administrative budget. The portion of the bar that falls below the 0 base line would show the deficit portion of the budget (Fig. 160).

FIG. 156

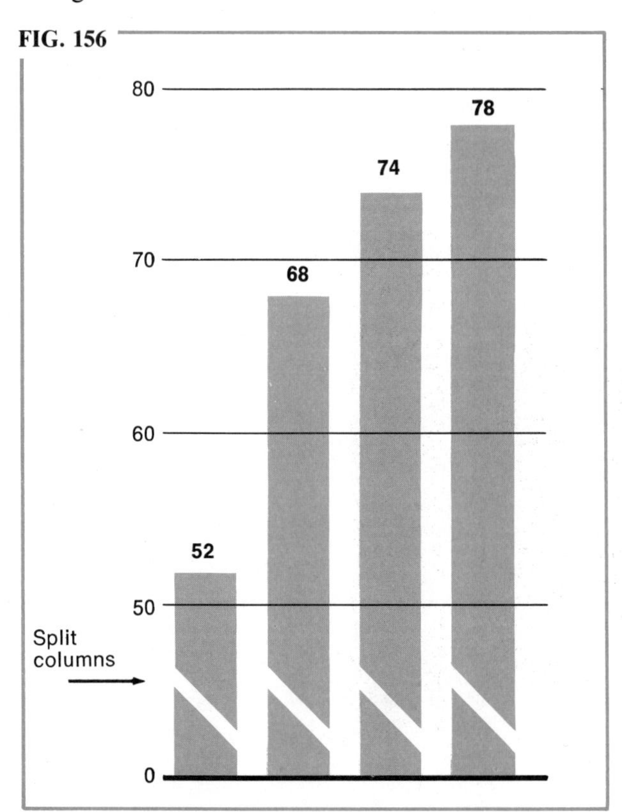

FIG. 157

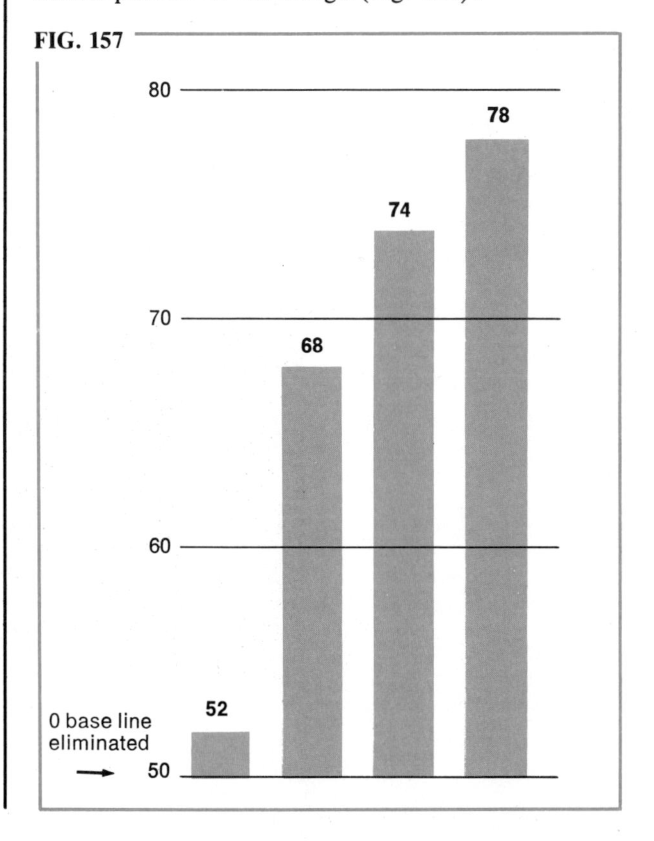

FIG. 158

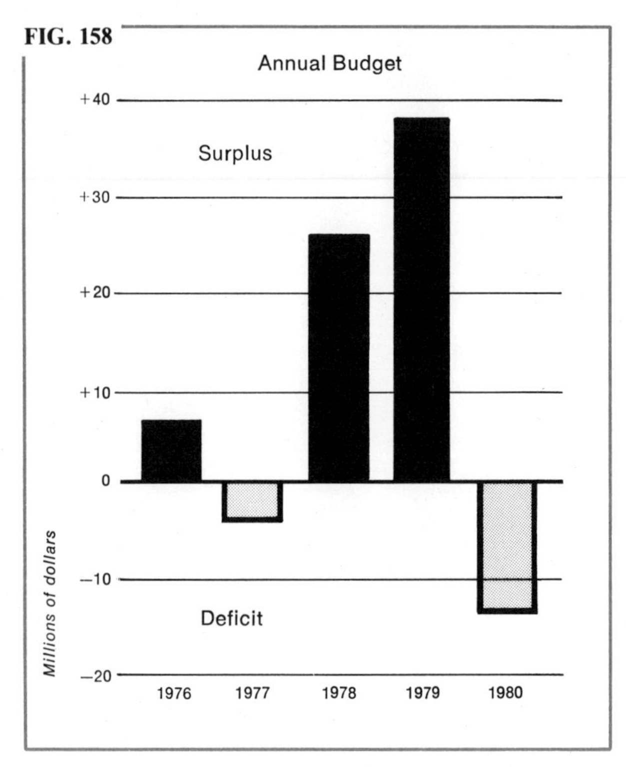

FIG. 159

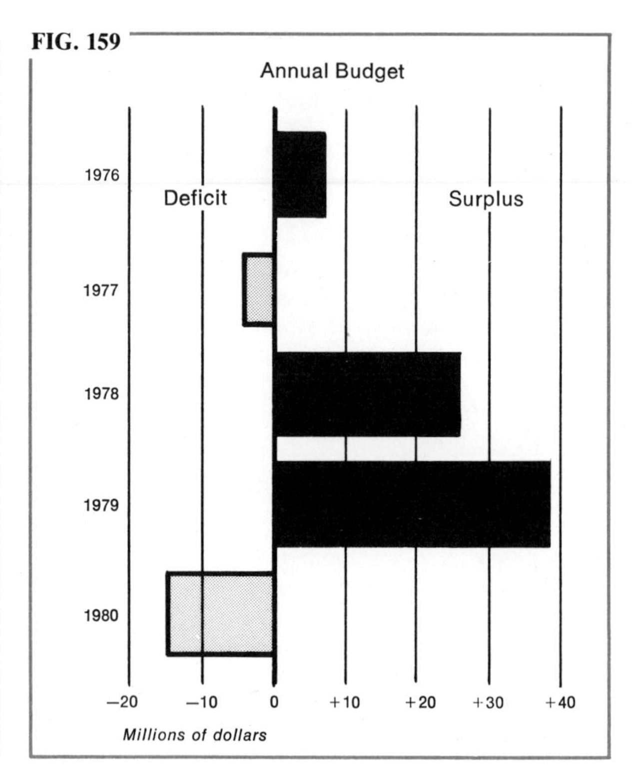

FIG. 160

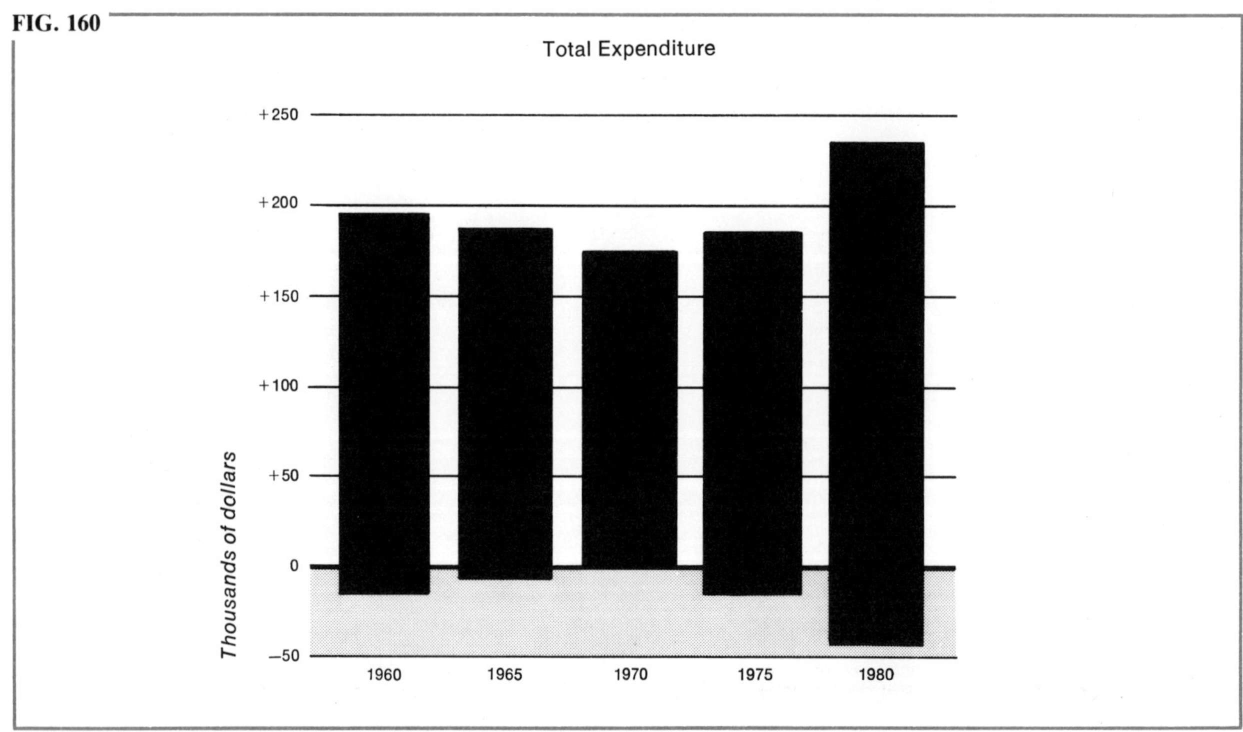

MULTIPLE SCALE

This type of chart is used to compare two elements that require different units or different scales. The numbers might also be in the same unit values but have too broad a range to be reasonably illustrated in one chart. For example, a comparison between Net Sales and Net Earnings might be the following:

In Millions of Dollars

	Net sales	Net earnings		Net sales	Net earnings
1970	$265	$3.2	1976	312	2.8
1971	282	2.8	1977	448	8.2
1972	315	2.4	1978	660	12.1
1973	360	5.0	1979	692	11.8
1974	375	3.0	1980	898	17.1
1975	261	3.0			

FIG. 161

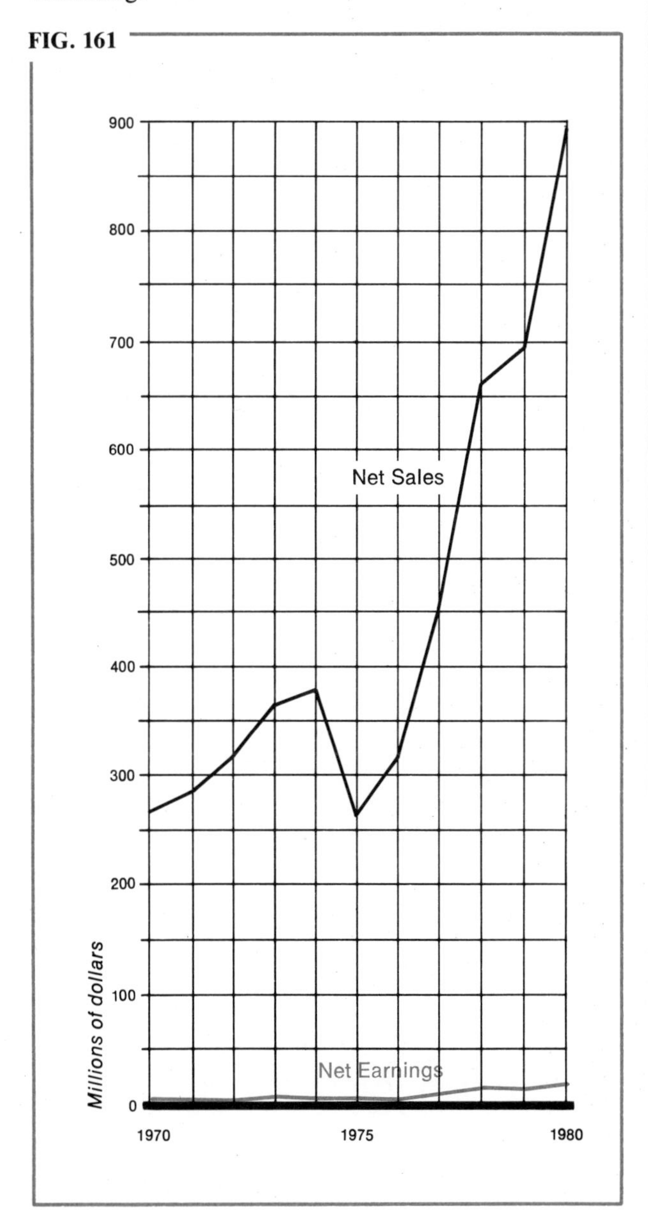

FIG. 162

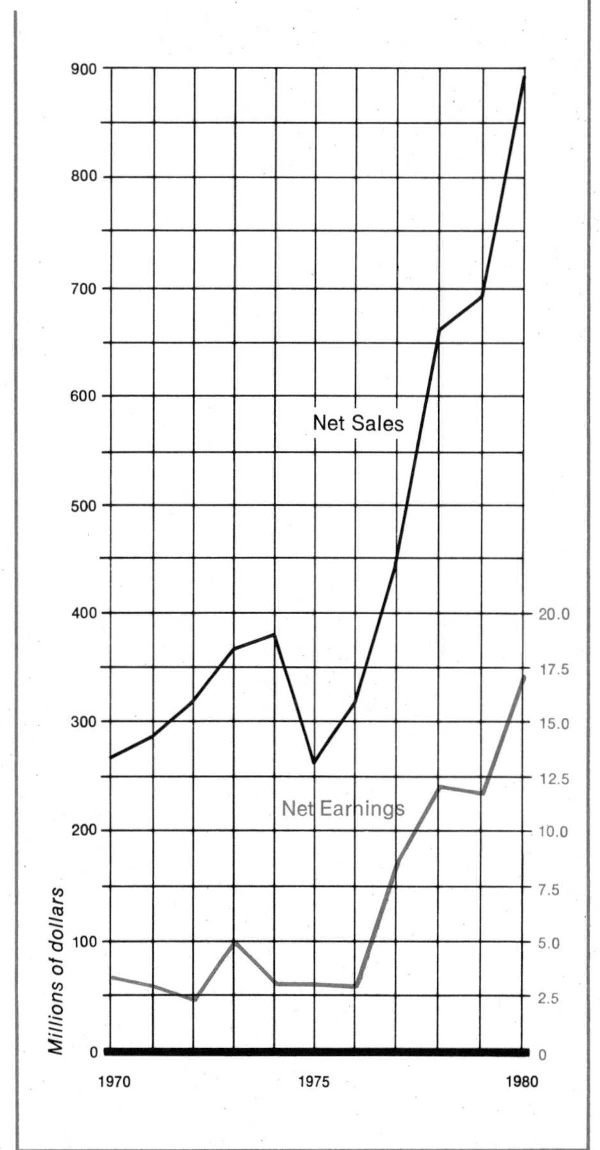

Both are in units of millions of dollars. The lowest figure is 2,400,000 and the highest figure is $898,000,000. A very broad jump. If drawn within a given area this chart would look like Fig. 161.

Note that it is impossible to plot the figures for Net Earnings with any degree of accuracy using the same scale as for Net Sales.

In order to accommodate both sets of extreme values, they can be treated as though they were separate charts. However, by using a common base line and one scale with two sets of values we can illustrate both sets of components with equal emphasis. Using both sides of the scale, we can establish the left range from $0 to $900,000,000 in increments of $50,000,000, whereas the right side of the same increments start as $0 and end at $20,000,000, each increment being only $2,500,000. Using a different color for each component will help identify them and enhance readability. Quite an effective method (Fig. 162).

Chapter Eight
Never a Dull Chart

THE PICTOGRAPHIC CHART

Charts in general present cold facts. They "tell it like it is." They can be dressed up to stimulate interest and assist comprehension. Among the style variations in graphic presentation of charts, the use of symbols or pictures is one of the most prominent.

Particularly applicable to the column or bar chart, symbols add a certain depth and explicit description to its meaning (Fig. 163).

My objective has been to teach you how to *prepare* charts, not how to design them. Nevertheless, you should be acquainted with some basic graphic devices used to dress up a chart. The following pages display a variety of elementary approaches in style. Their purpose is to show that charts need not be cold and drab. Of course, the more you have to work with, the more you can do with a chart.

FIG. 163

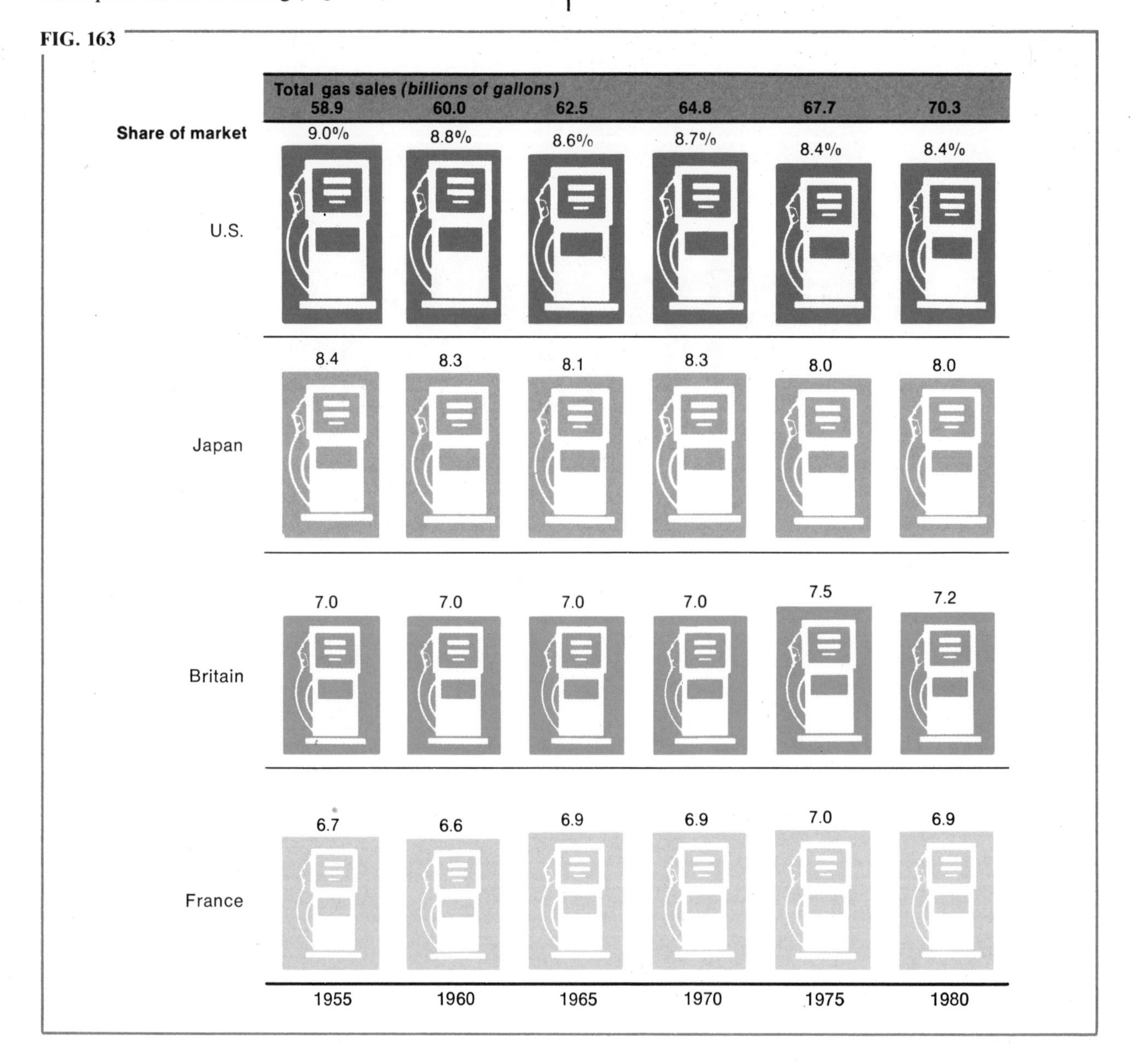

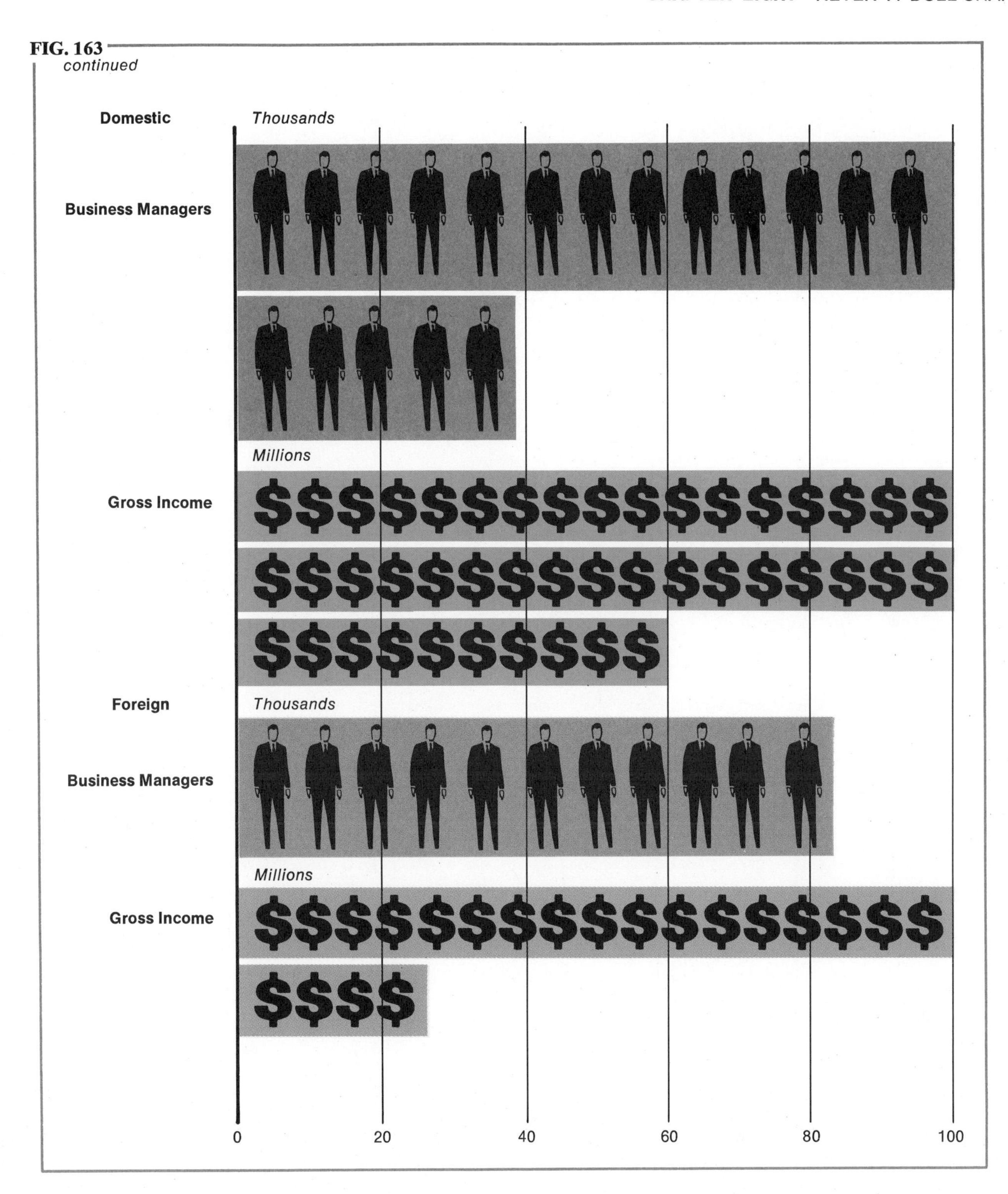
FIG. 163
continued
Domestic
Thousands
Business Managers
Millions
Gross Income
Foreign
Thousands
Business Managers
Millions
Gross Income
0
20
40
60
80
100

LINE WEIGHT COMBINATIONS

Figure 164 is the basic form for a typical column or bar chart. While it depicts the relationship of one figure to the other, it is done in a matter-of-fact manner. One might look at it and say, "So what?" (Fig. 164).

FIG. 164

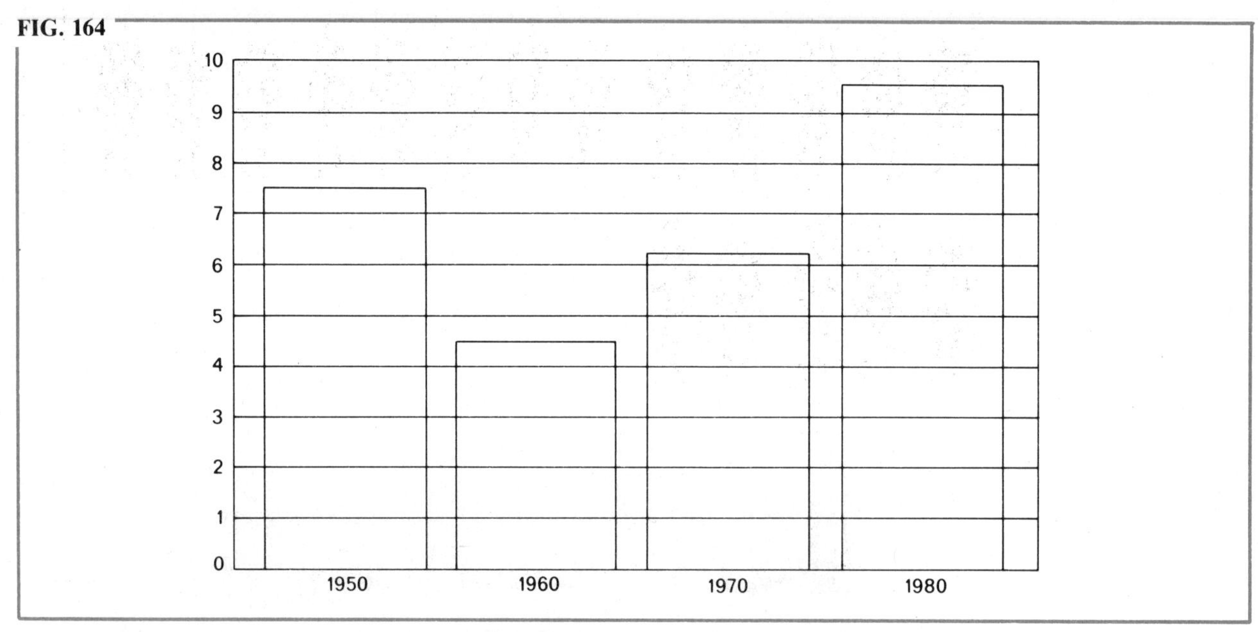

However, by simply varying the line weight combinations you will find that the same chart begins to take on an air of seriousness (Fig. 165).

FIG. 165

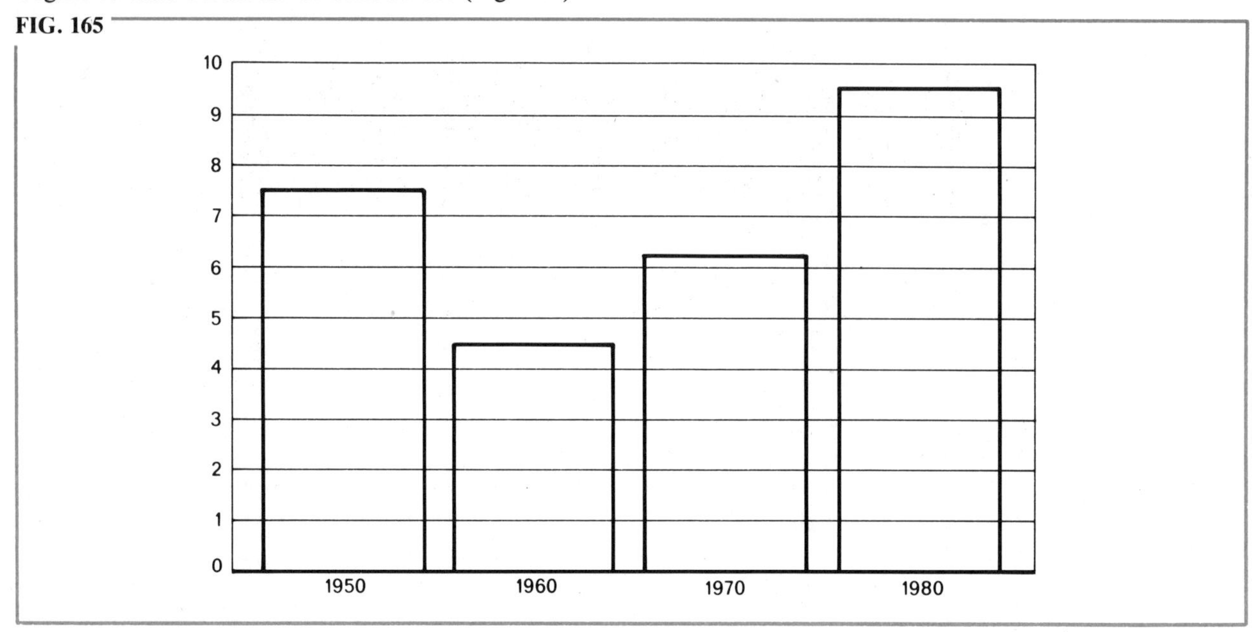

If you add a third weight and stronger contrast, it certainly looks precise (Fig. 166).

FIG. 166

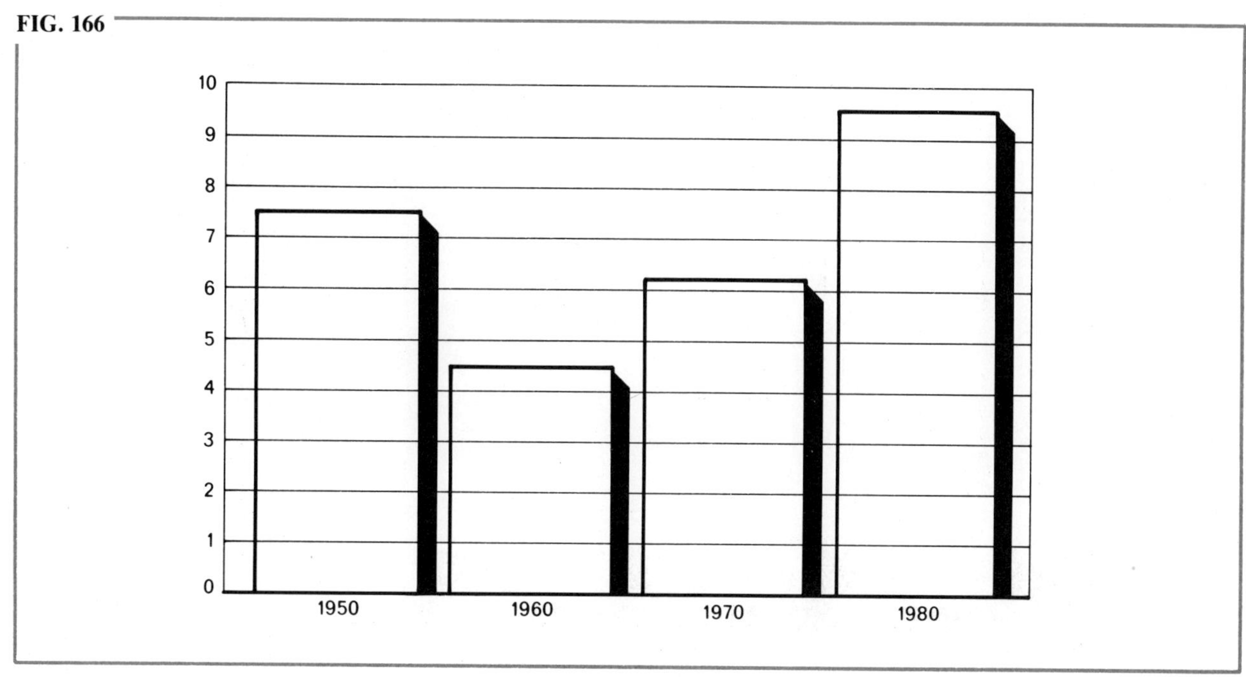

Eliminating unnecessary lines gives the chart added dimension (Fig. 167).

FIG. 167

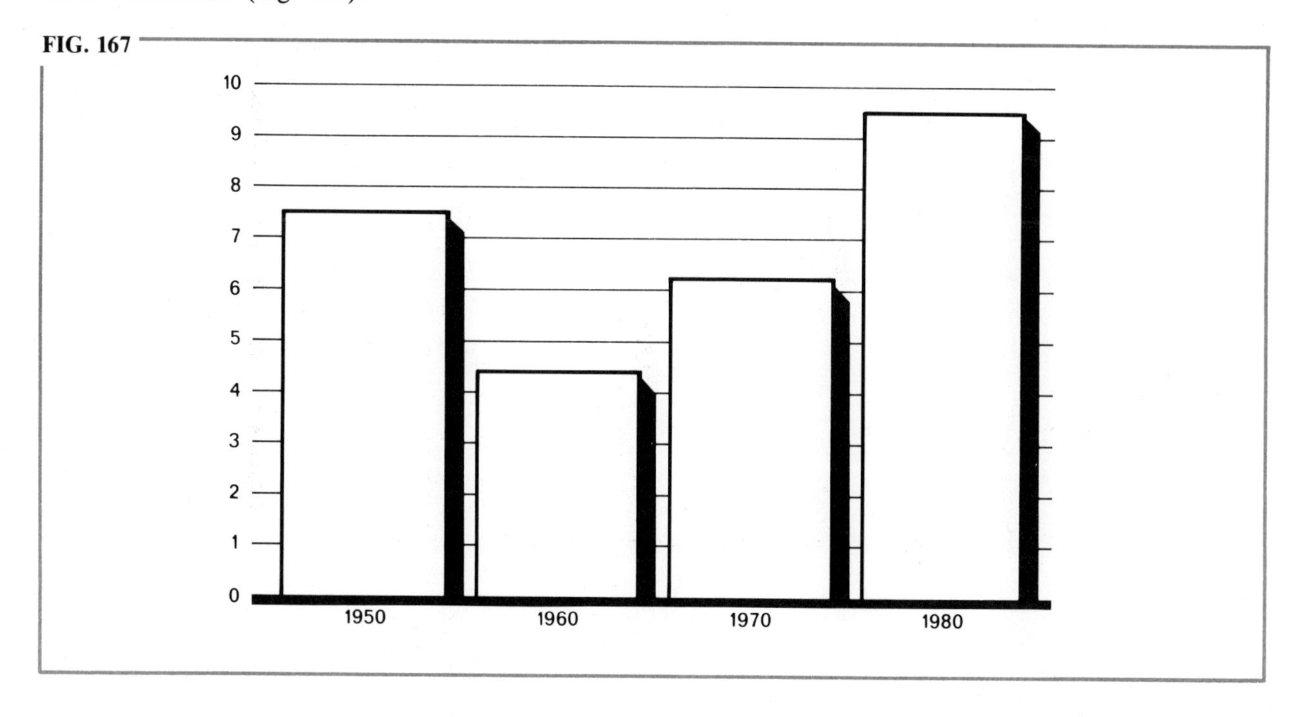

Add tone or color or a texture, and your chart comes alive (Fig. 168).

FIG. 168

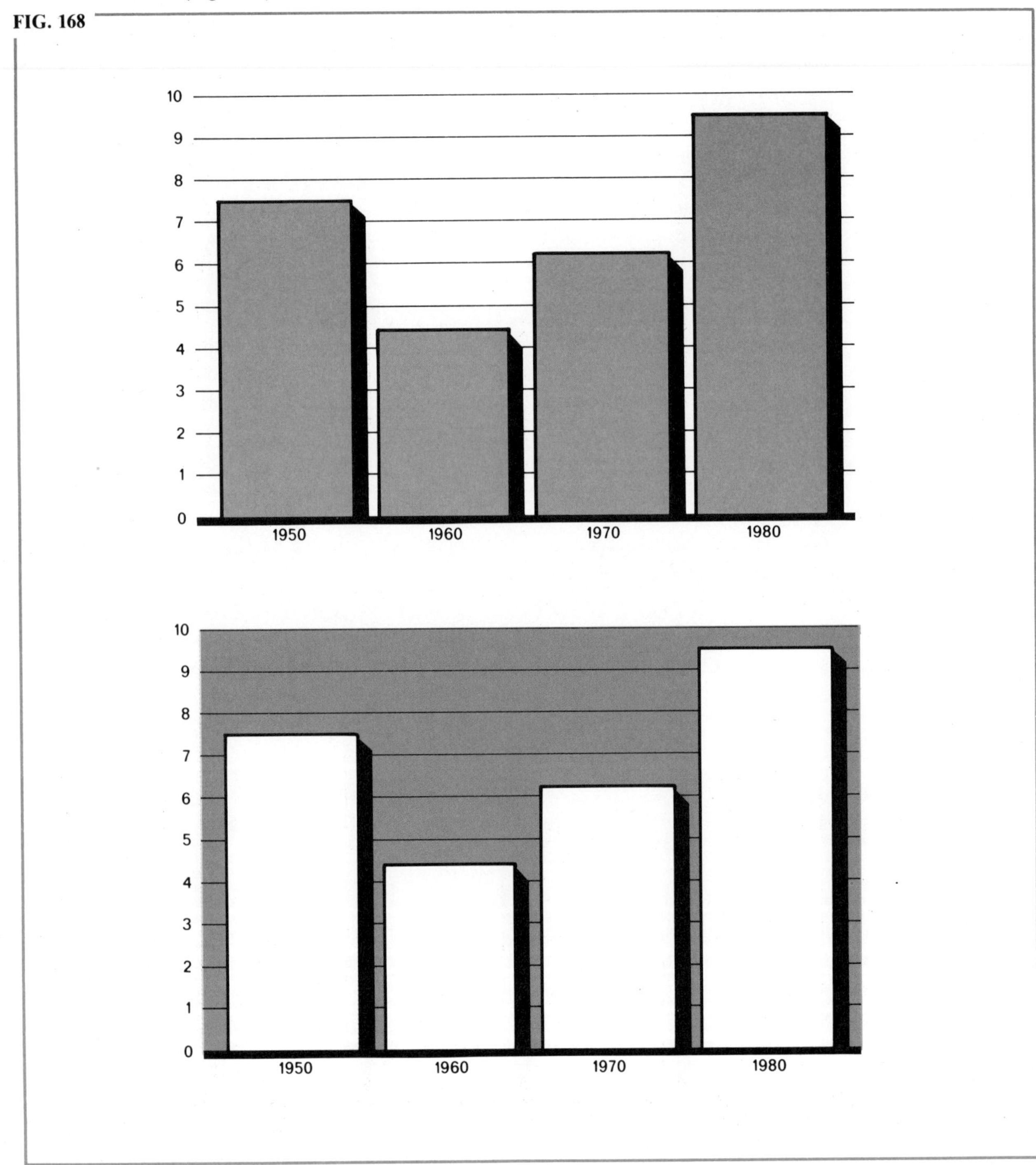

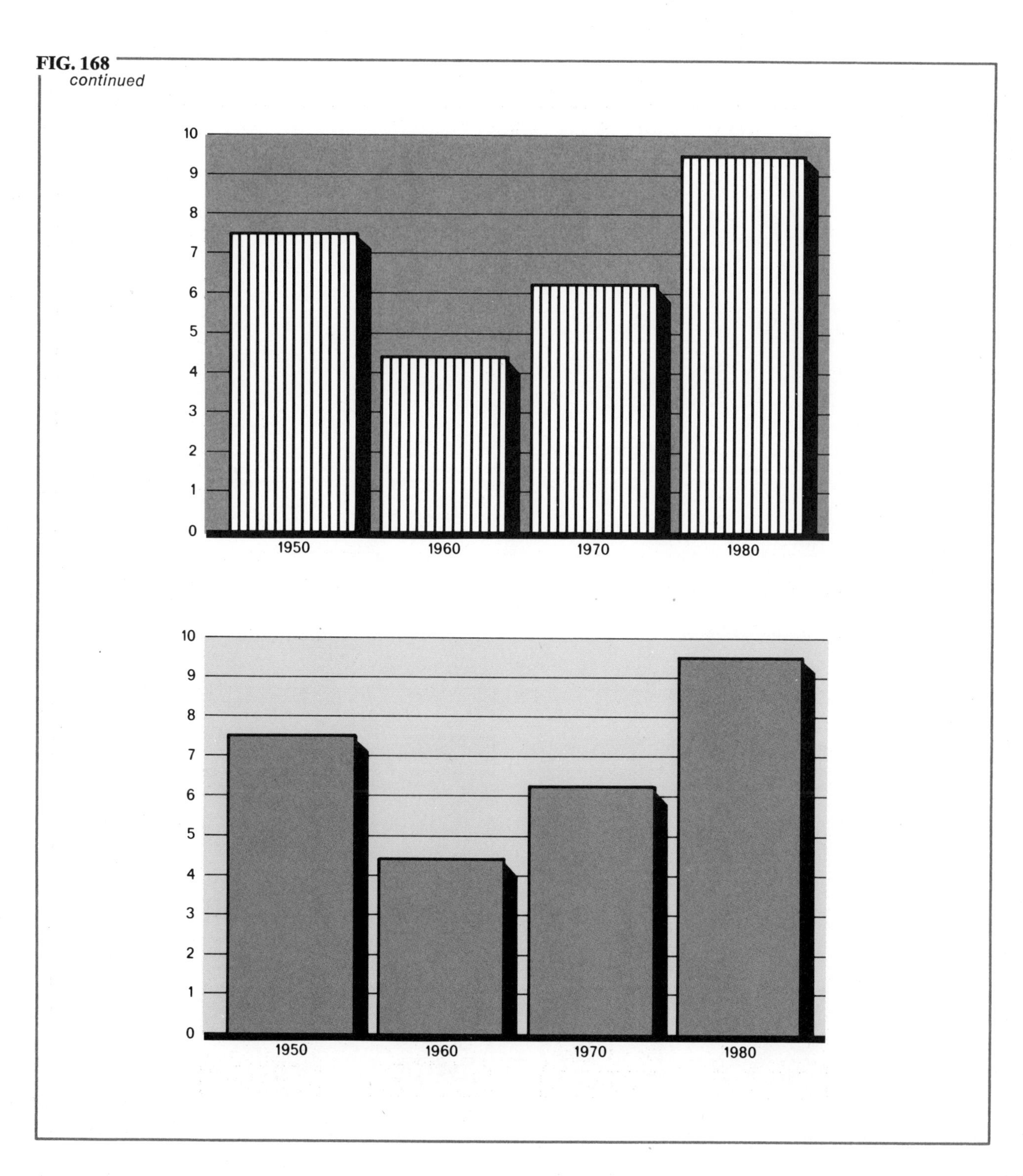
FIG. 168
continued
10
9
8
7
6
5
4
3
2
1
0
1950
1960
1970
1980
10
9
8
7
6
5
4
3
2
1
0
1950
1960
1970
1980

A SUBDIVIDED COLUMN CHART

This type of chart is quite suitable to variation. Once again, the basic delineation of the facts could be rather bland (Fig. 169).

FIG. 169

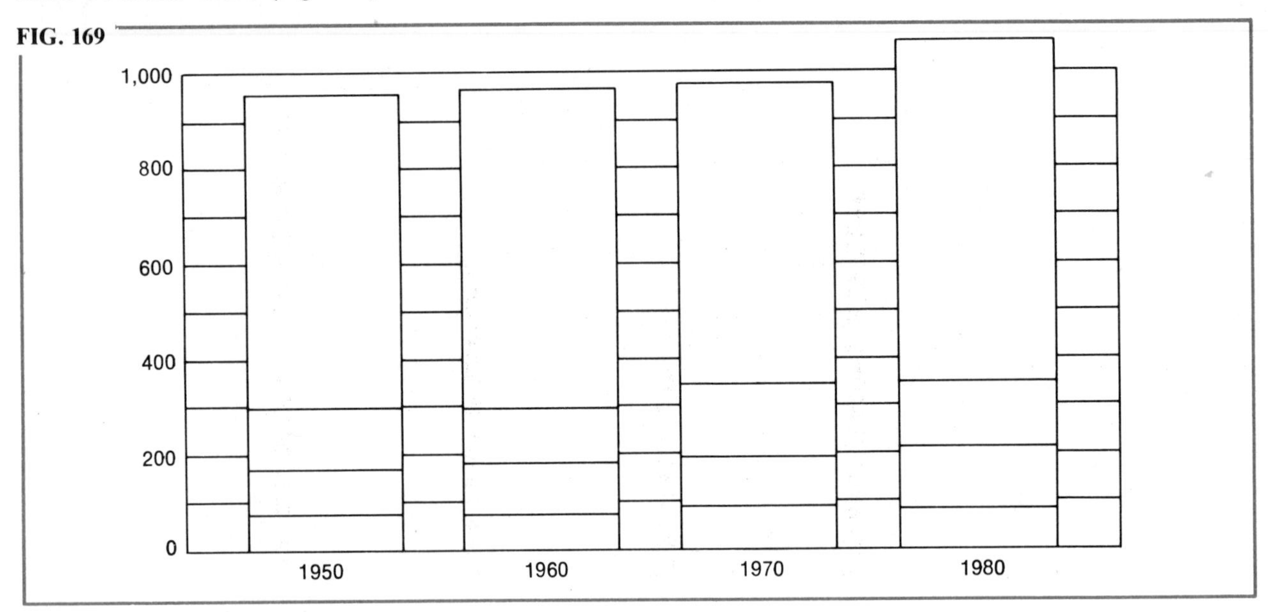

But when you begin to dress it up, it looks like a different chart. It isn't different at all. The facts are the same; we have just added modulation to our statement (Fig. 170).

FIG. 170

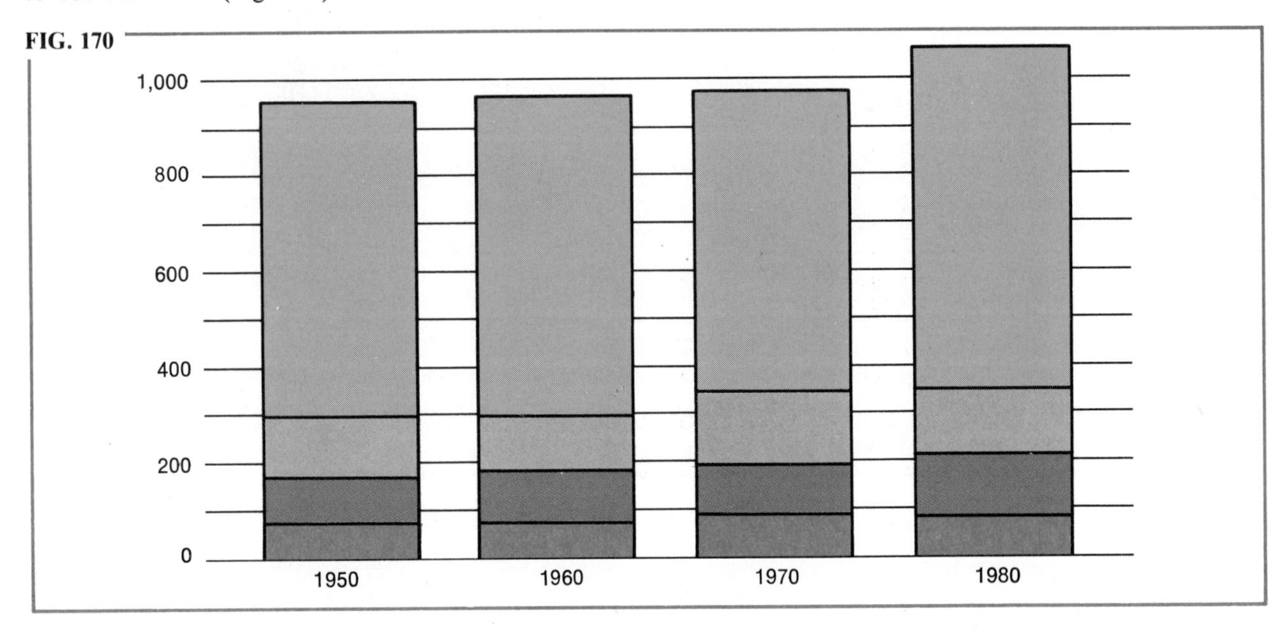

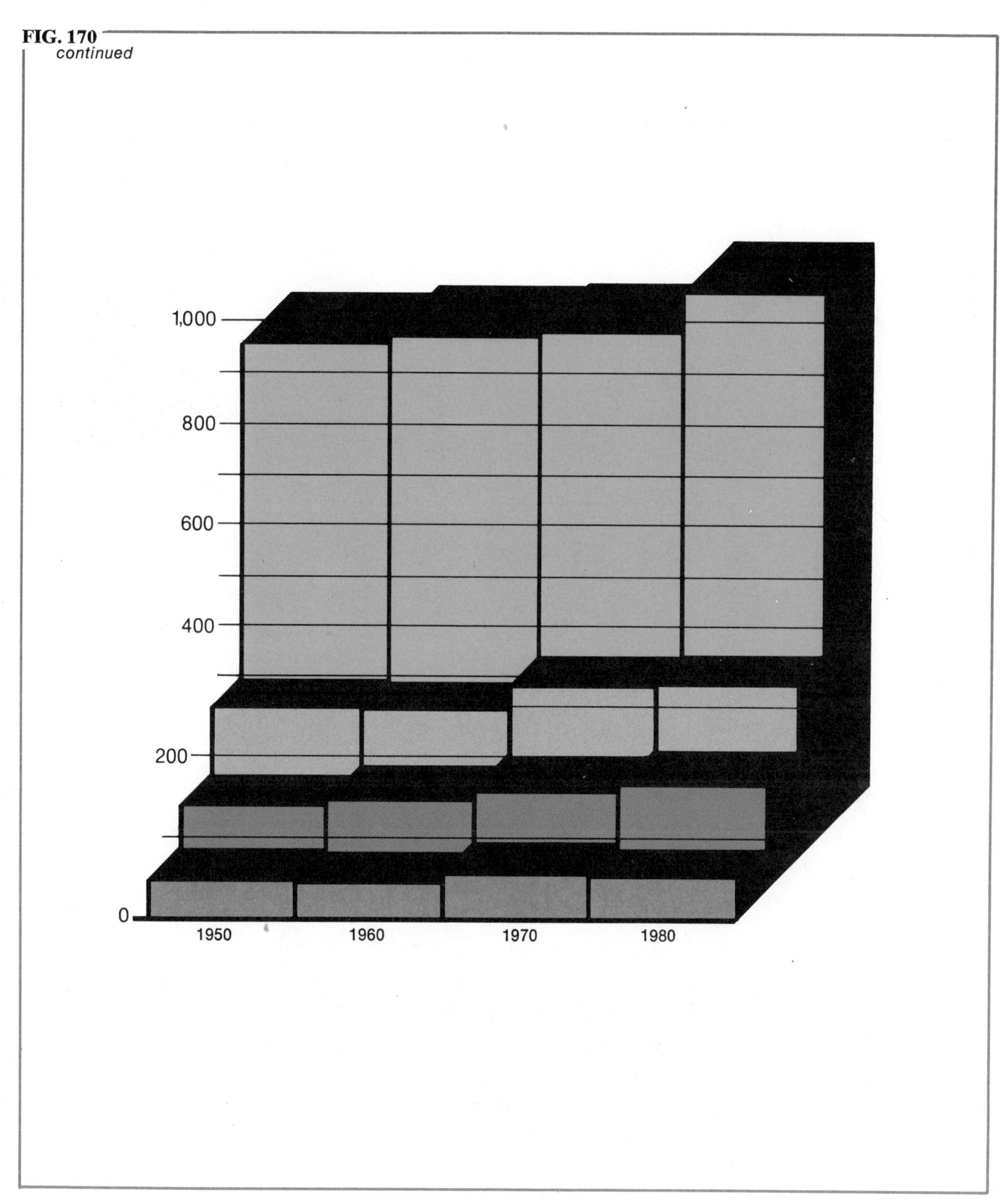
FIG. 170
continued
1,000
800
600
400
200
0
1950
1960
1970
1980

When one looks at a chart, it should explain itself in silence; it should be completely understood without the assistance of a caption. The caption must act only as a reinforcement to its comprehension. This is a challenge, of course, but it is what makes a chart good (Fig. 171).

FIG. 171

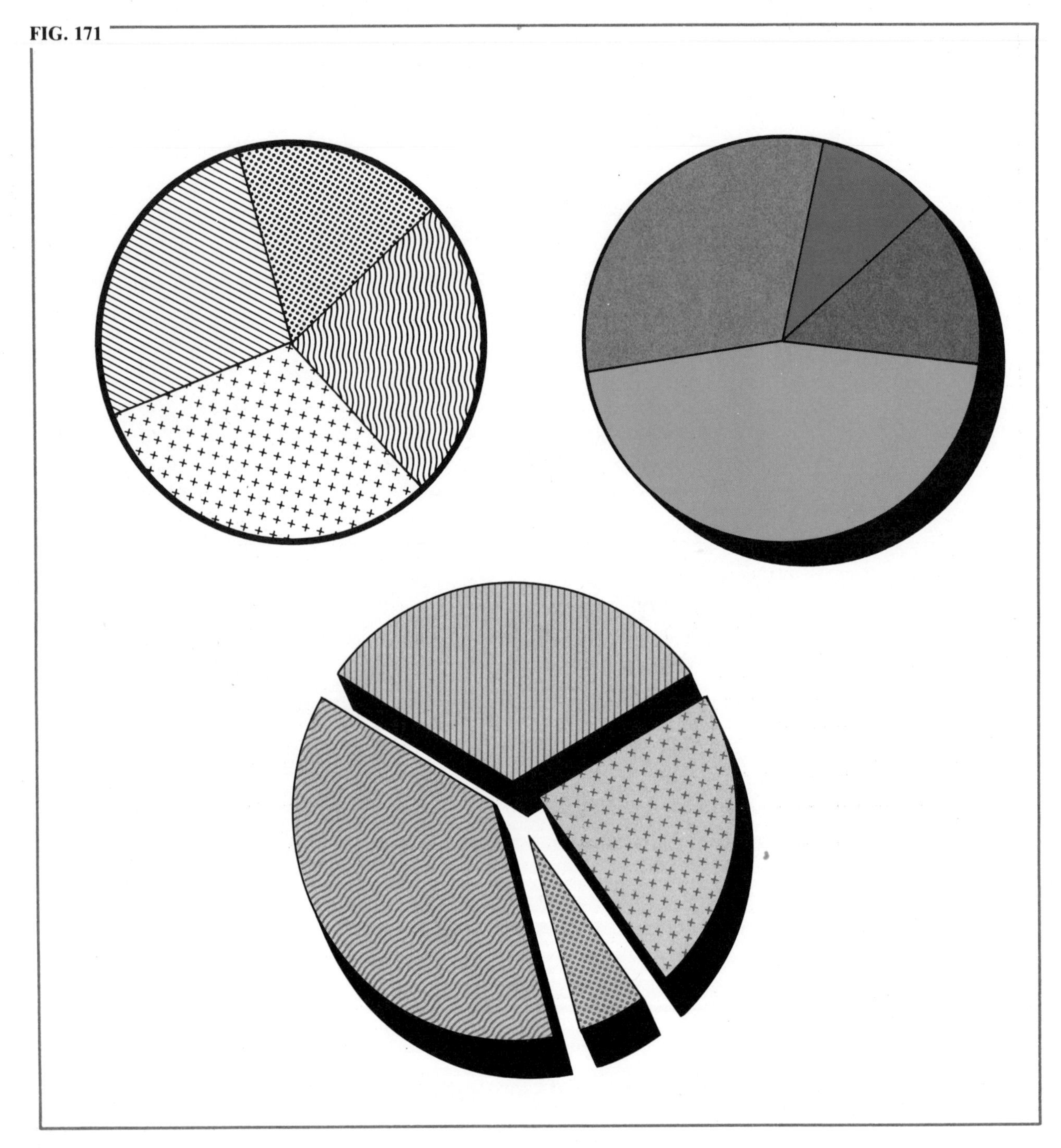

FIG. 171
continued

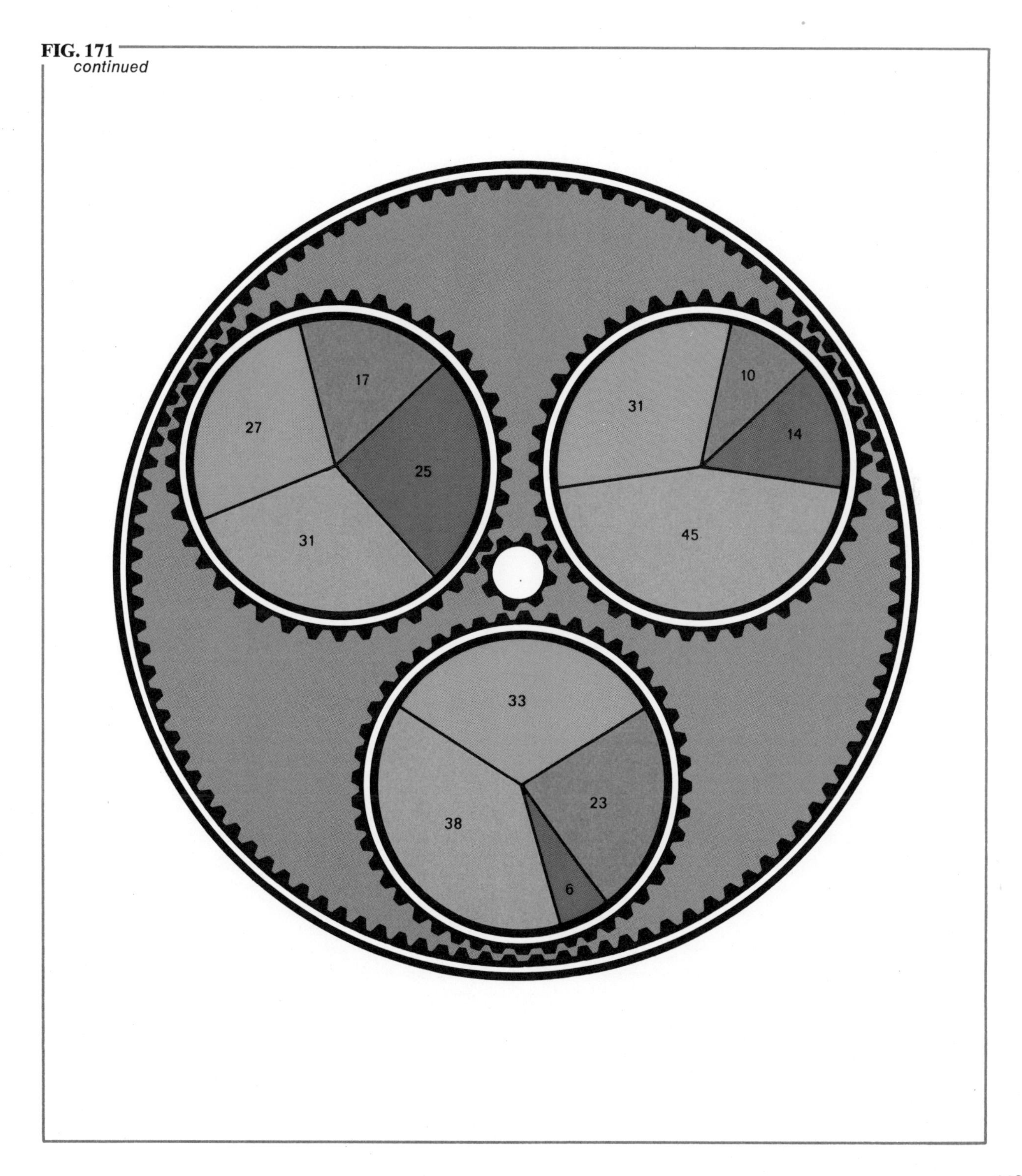

The combinations are endless (Fig. 172).

FIG. 172

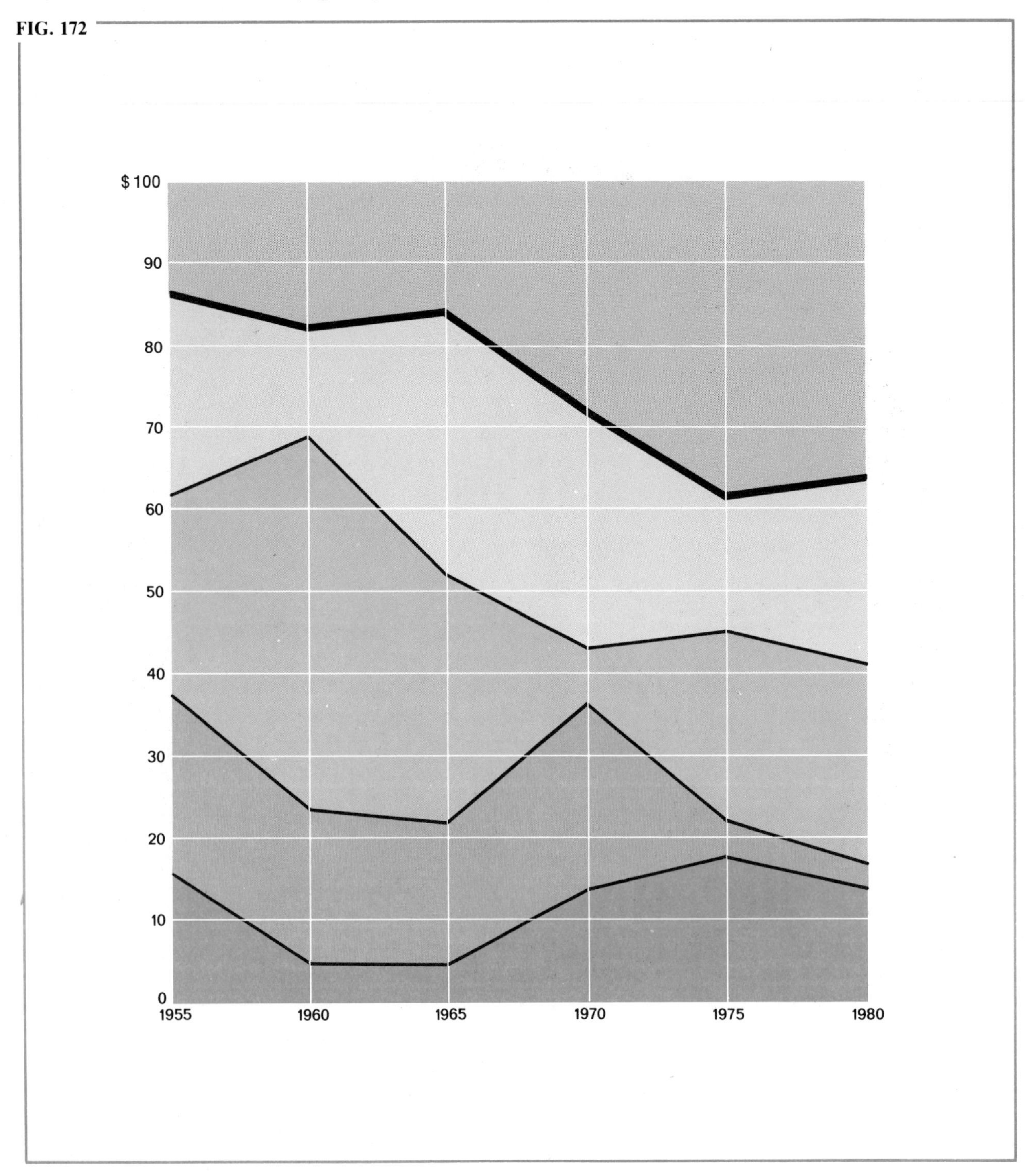

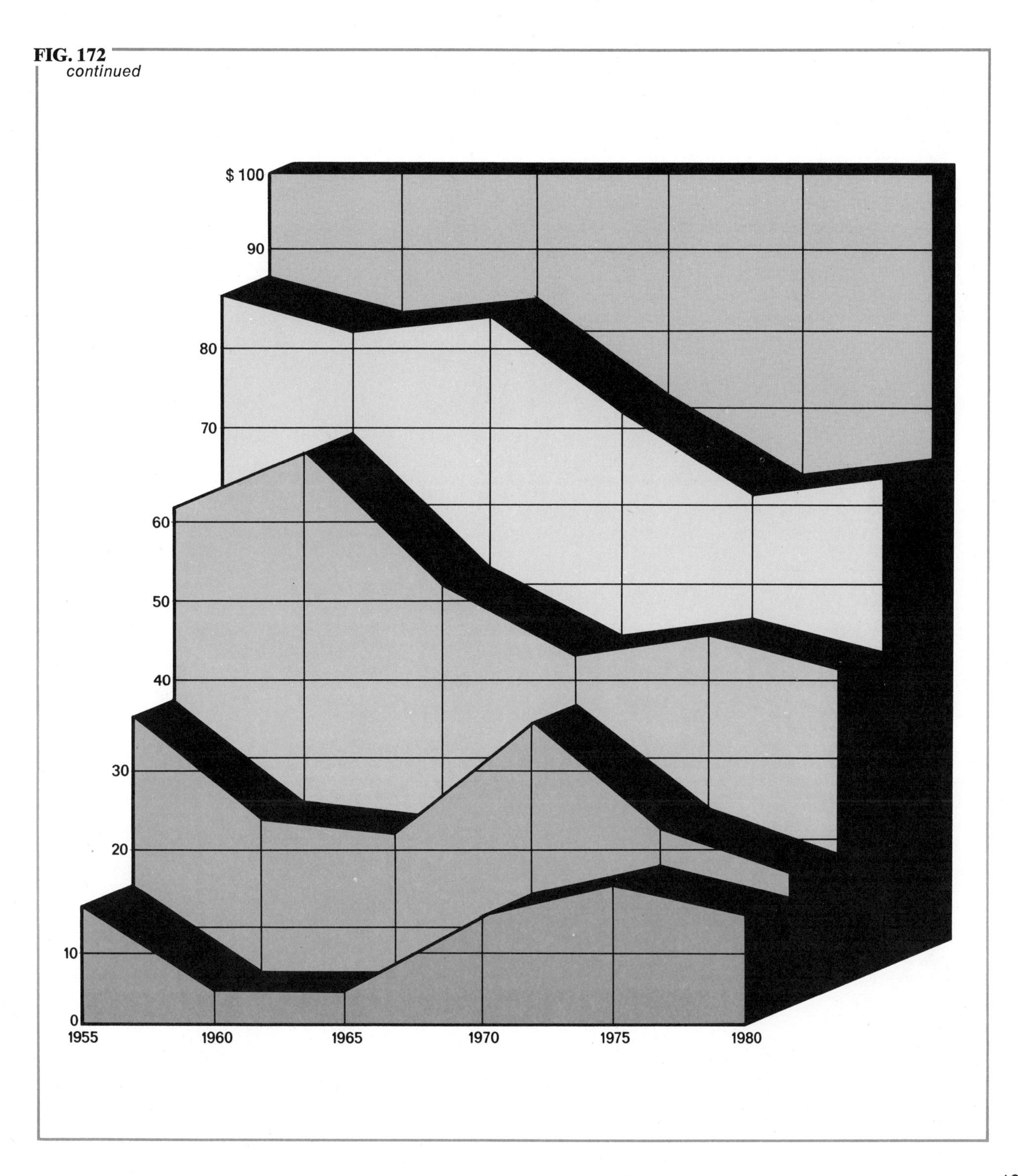
FIG. 172
continued
$ 100
90
80
70
60
50
40
30
20
10
0
1955
1960
1965
1970
1975
1980

Once you begin to exercise your imagination, you will discover that charts can be made quite interesting. Remember, the purpose of a chart is to express and convey a message. A visual graphic presentation is essentially the same as an audiovisual presentation. It can be stated flatly and without dramatization, or it can be done with excitement and style. Just as in making a speech one can stress a particular point with a change of voice or pitch or an added gesture, so can a chart be enhanced by the use of color, contrast of elements, and the use of graphic devices.

You know now how to construct the foundation and framework of a chart. With practice and ingenuity you will soon be able to give it personality. Good luck!

Answers to Exercises

Exercise, page 54

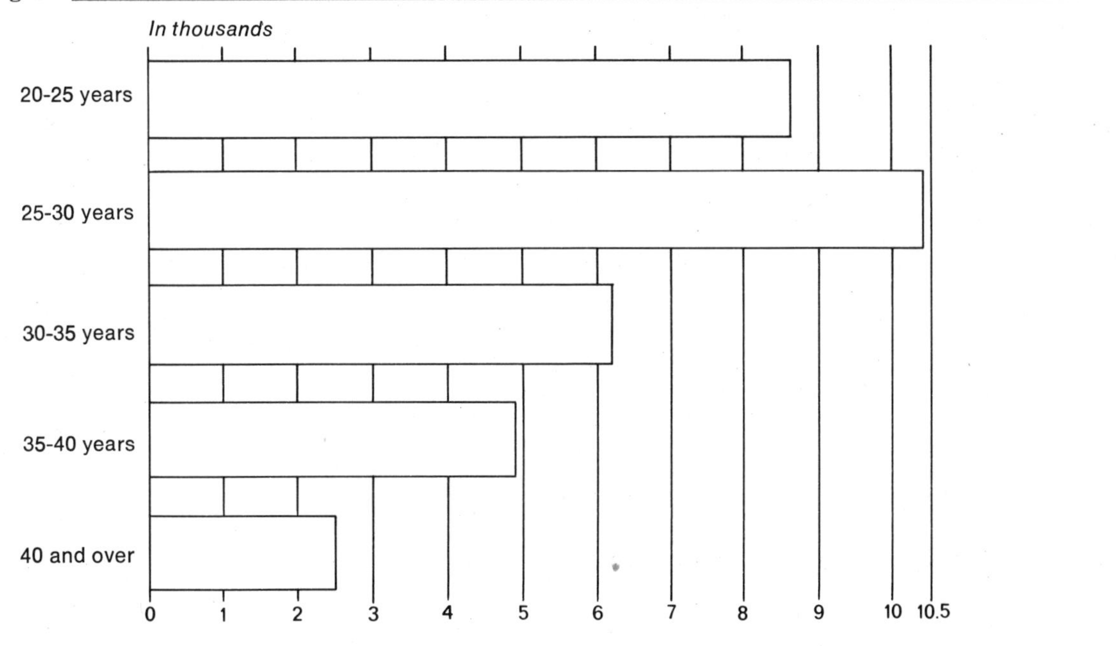

Exercise, page 60

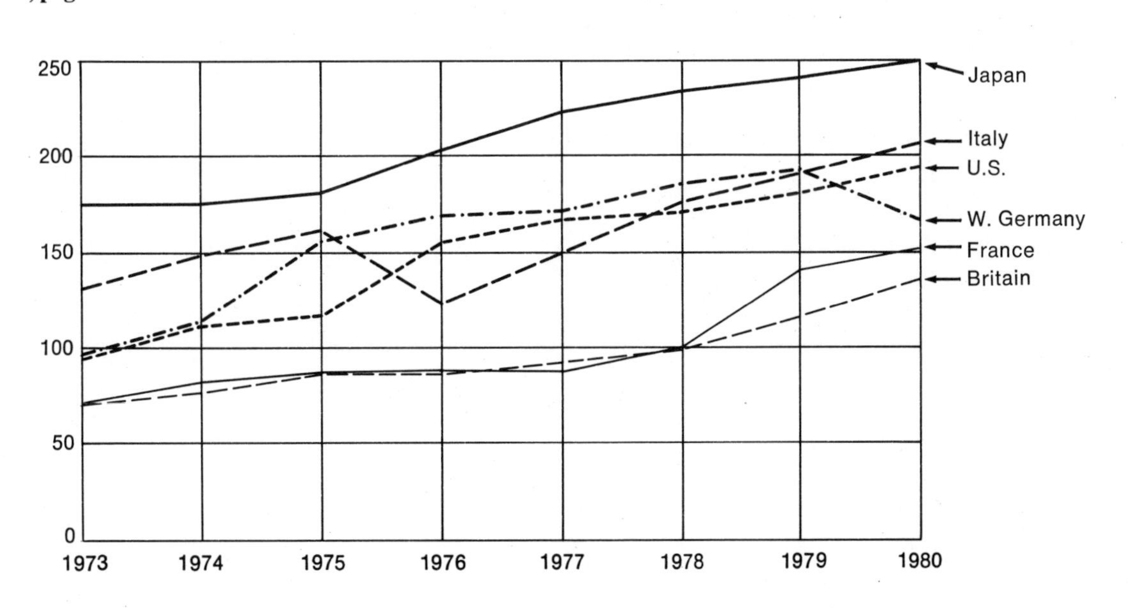

Answers to Exercises *continued*

Exercise, page 68

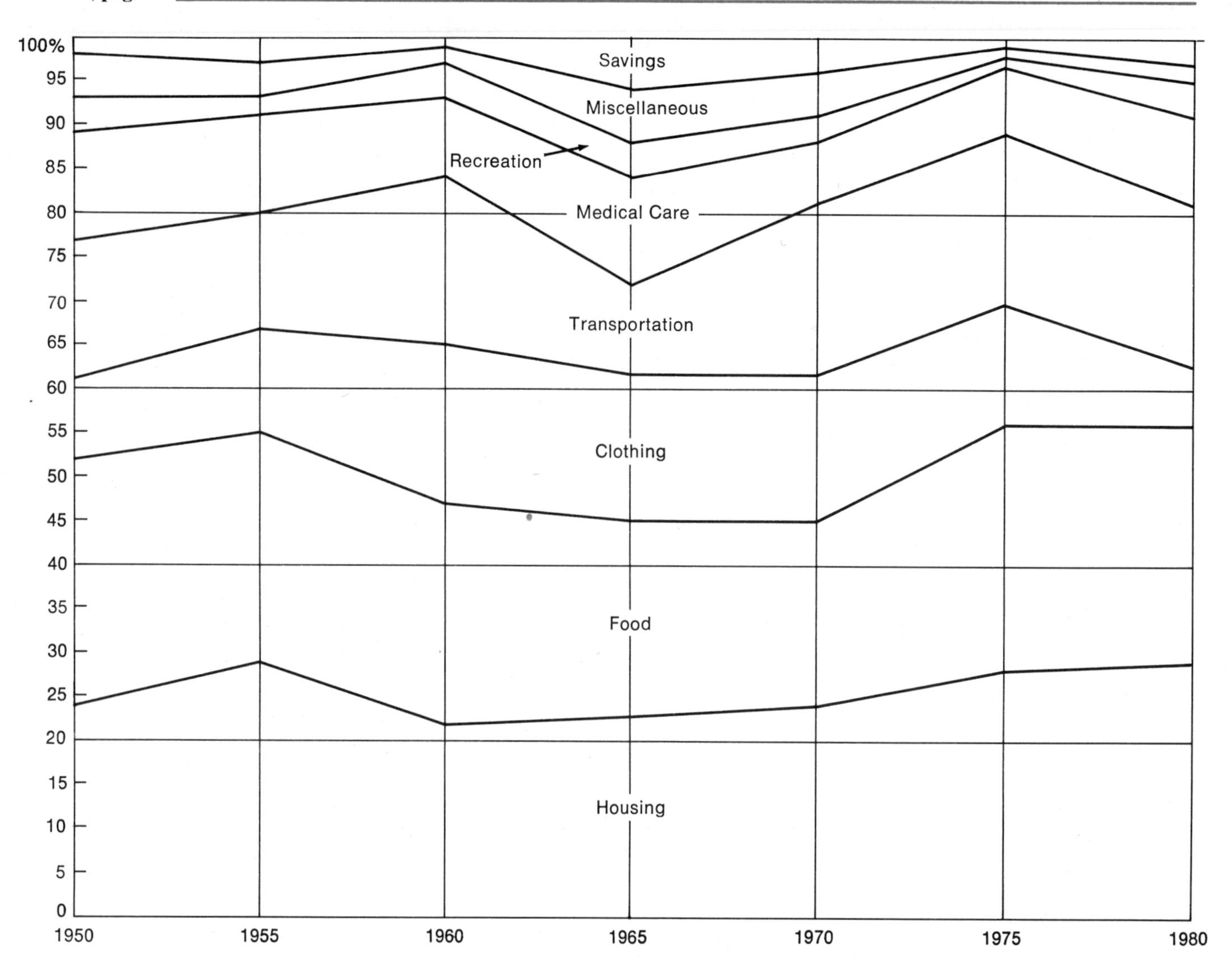

Answers to Exercises *continued*

Exercise, page 84

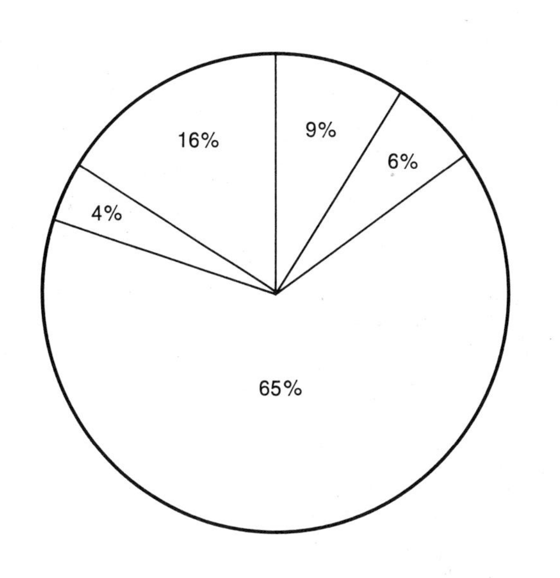

Exercise A, page 100

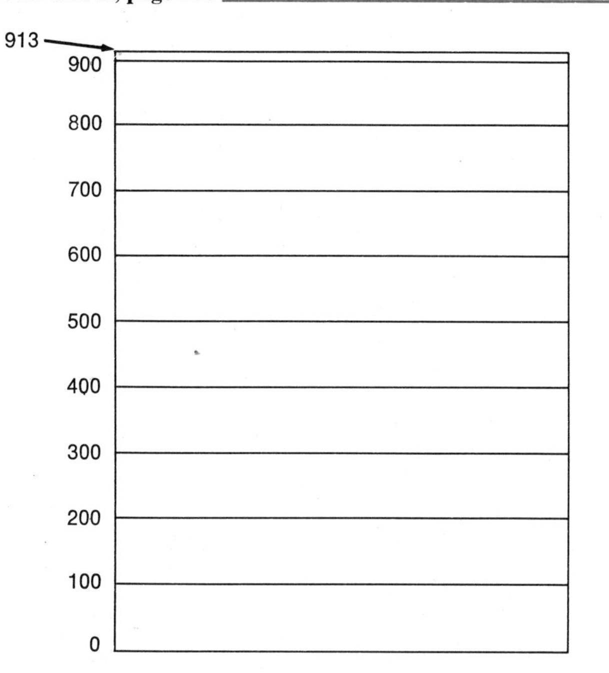

Exercise B, page 100

Exercise, page 104

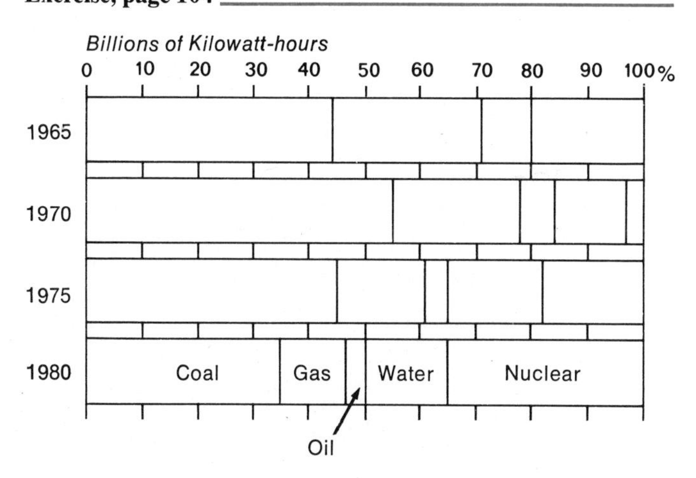

Here are the converted percentage figures:

1965	44%	27%	9%	20%	0%
1970	55%	23%	6%	13%	3%
1975	45%	16%	4%	17%	18%
1980	35%	12%	3%	15%	35%

The Pie Scale

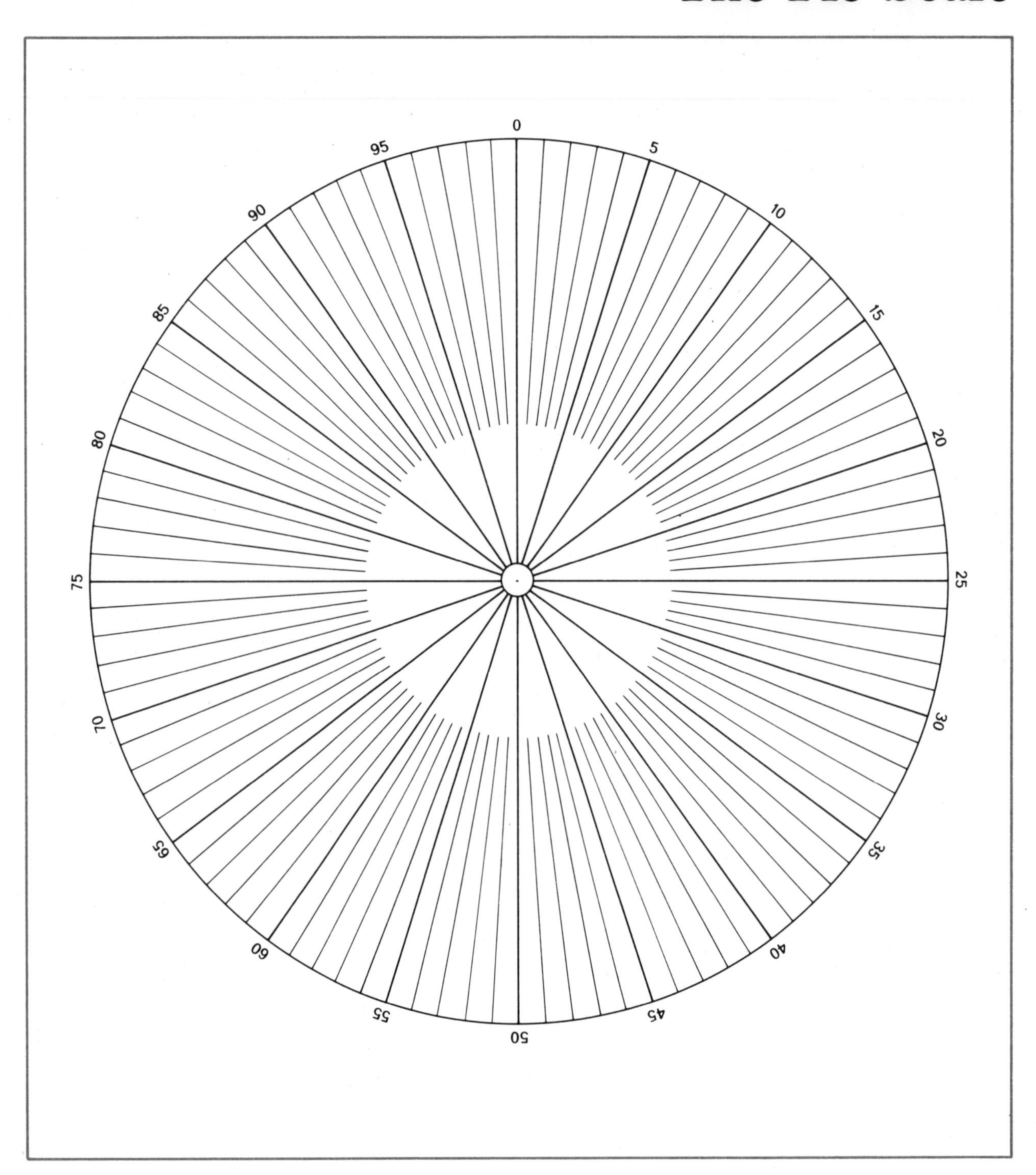

Index

4/2

Designed by Ann Kahaner
A special note of thanks to Ted Plonchak and Bonnie Cook
Editorial assistance by Doreen Kalter